AF614764

Modern Applications of Wavelet Transform

Edited by Srinivasan Ramakrishnan

Published in London, United Kingdom

Modern Applications of Wavelet Transform
http://dx.doi.org/10.5772/intechopen.1000342
Edited by Srinivasan Ramakrishnan

Contributors
Srinivasan Ramakrishnan, Bianca-Gabriela Antofie, Marius-Ciprian Larco, Teodor Lucian Grigorie, Hassina Boukerbout, Okonkwo Chidi Ukwuoma, Ugo Donald Chukwuma, Titus Ifeanyi Chinebu, Maysoun Alshrouf, Cajetan M. Akujuobi, Emad Awada

First published in London, United Kingdom, 2024 by IntechOpen
IntechOpen is the global imprint of INTECHOPEN LIMITED, registered in England and Wales, registration number: 11086078, 5 Princes Gate Court, London, SW7 2QJ, United Kingdom

British Library Cataloguing-in-Publication Data
A catalogue record for this book is available from the British Library

Additional hard and PDF copies can be obtained from orders@intechopen.com

Modern Applications of Wavelet Transform
Edited by Srinivasan Ramakrishnan
p. cm.
Print ISBN 978-0-85466-236-4
Online ISBN 978-0-85466-235-7
eBook (PDF) ISBN 978-0-85466-237-1

For EU product safety concerns:
IN TECH d.o.o., Prolaz Marije Krucifikse Kozulić 3, 51000 Rijeka, Croatia,
info@intechopen.com or visit our website at intechopen.com.

Meet the editor

Dr. Srinivasan Ramakrishnan has twenty-three years of teaching experience and one year of industry experience. He holds a Ph.D. in Information and Communication Engineering from Anna University, India. He is the Dean of Research and Innovation at Dr. Mahalingam College of Engineering and Technology, India. Dr. Ramakrishnan is an associate editor for IEEE Access, a reviewer of forty-three international journals, and a guest editor of four special journal issues He has filed three Indian patents and published 216 papers in international and national journals. He has also published thirteen books on image processing, pattern recognition, soft computing, information security, and wireless sensor networks. Dr. Ramakrishnan has successfully guided nine Ph.D. scholars and is currently supervising five more. He was recognized as a top scientist in the AD Scientific Index in 2022–2023.

Contents

Preface

The wavelet transform has become a prominent and influential tool in signal processing for the analysis and representation of signals. It provides a distinctive and valuable method for analyzing and representing signals. This mathematical technique, which has the potential to bring about enormous advances, has been applied in a range of sectors, including the processing of pictures and sounds, communication systems, and other areas. An extensive examination is underway to explore the intricate realm of wavelet transform applications. This book provides a comprehensive overview of wavelet transformation and describes its contemporary applications. This book is comprised of five captivating chapters.

Chapter 1 introduces the basic principles of wavelet transformations, providing a solid foundation for further exploration. As an introductory guide, the chapter's objective is to familiarize readers with the essential concepts that form the basis of this mathematical framework. This chapter establishes the essential foundation for the subsequent discussions by analyzing both continuous and discrete wavelet transforms. This chapter provides a vast list of applications of wavelet transform.

Chapter 2 investigates the interaction between tiny inertial sensors and wavelet filtering techniques, showing the potential for increased denoising performance by collecting localized and non-stationary oscillations. This chapter additionally examines the utilization of wavelet filters in MEMS-based inertial sensors, showcasing substantial enhancements in altitude parameters and location precision.

Chapter 3 explores the application of intricate wavelet and ridgelet transforms in geophysical prospecting. It discusses how these transforms are employed to analyze probable field data and detect hidden structures that are accountable for anomalies. This chapter employs these transformations to detect the origins of anomalies in potential fields in both two-dimensional and three-dimensional scenarios. Additionally, it explores the utilization of artificial intelligence for grouping Euler solutions. The chapter offers useful insights into the interpretation of potential fields in geophysical data.

Chapter 4 delves into wavelet analysis as a mathematical instrument for identifying volatility patterns in asset returns. It dissects financial time series data into several scales and frequencies, offering a comprehensive view of both short-term and long-term patterns. This chapter presents applications of wavelet in the field of finance as it is utilized through many R packages. This chapter examines the stock price returns of NASDAQ Composite, DOW Incorporated, S&P500, and Omnicell Inc. It reveals that there is significant volatility at lower frequencies, but lower volatility at higher frequencies.

Chapter 5 presents a fault detection technique that utilizes the DWT to evaluate and identify short-circuit faults in a three-phase power system transmission line. It presents an algorithm that utilizes phase-to-phase and phase-to-ground fault analysis to accurately identify different types of faults. The objective of the chapter is to enhance the precision of power system monitoring and explore the benefits of DWT in identifying, localizing, and categorizing anomalous failures in a three-phase power system short circuit. The authors employ MATLAB/Simulink to simulate various fault scenarios on transmission lines, extracting intricate coefficients and comparing them against a predefined threshold.

This book is specifically designed to meet the needs of a wide range of readers, including students, researchers, and professionals who have a keen interest in gaining a comprehensive knowledge of wavelet transforms and their practical implementations. Every chapter has been meticulously crafted to offer a blend of theoretical principles and practical examples, ensuring a thorough and easily understandable exploration of the subject matter.

I would like to express my sincere thanks to all the authors for their contributions and efforts to bring about this wonderful book. My earnest gratitude and appreciation go to the staff at IntechOpen, particularly Publishing Process Manager Ms. Laura Divic, who brought together the authors to publish this book. I would like to express my heartfelt thanks to the management, secretary, and principal of my institute. Finally, dearest thanks to my family members, especially my sweet daughter Abirami.

Dr. Srinivasan Ramakrishnan
Dean of Research and Innovation,
Dr. Mahalingam College of Engineering and Technology,
Pollachi, India

Chapter 1

Introductory Chapter: Wavelet Theory and Modern Applications

Srinivasan Ramakrishnan

1. Introduction of wavelet transform

The wavelet transform is a mathematical technique used to analyze images and signals. It decomposes them into discrete frequency components with different resolutions. This allows for localization in both time and frequency domains. Unlike the Fourier transform, which primarily focuses on frequency information, the wavelet transform captures data at both high and low frequencies within a signal [1–3].

The continuous wavelet transform (CWT) is a mathematical methodology used for continuous analysis of signals and data. It involves integrating a signal with shifted and rescaled iterations of a continuous mother wavelet function. The discrete wavelet transform (DWT) is another method that partitions a signal into distinct frequency components. However, it may not be computationally efficient compared to the CWT. The multiresolution analysis (MRA) is a mathematical methodology used to examine data at various degrees of resolution or intricacy [4].

The wavelet transform is used in various fields, including signal and image processing, compression algorithms, denoising, feature extraction, and biological signal analysis. Different wavelet families, such as Haar, Daubechies, Symlet, and Morlet, have unique attributes that make them suitable for specific applications [5, 6].

Wavelet transform is an adaptable and potent method that provides a thorough examination of signals and images at various scales, facilitating improved understanding and control of complex data in various disciplines [7].

2. Brief history of wavelet transform

Morlet's continuous wavelet transform was developed in the 1960s. Wavelets developed from the early work of Haar and Wiener. In the 1980s, Meyer, Daubechies, and Mallat made advancements in wavelet theory. Discrete transforms were introduced, and the theory was improved. In the 1990s, wavelets were used in applications like JPEG2000 and biomedical signal processing. Wavelets became essential in image compression, data analysis, and scientific research. They are now expanding their applications to include machine learning and big data analysis. Significant developments in wavelet transform have occurred every few years from 1966 to 2023 [8–15].

- Jean Morlet is the one who first presented the idea of continuous wavelet transform (CWT) in the field of geophysics in the year 1966.
- As seismic signal analysis and other geophysical applications continue to develop in the 1970s, wavelet principles continue to expand.
- Yves Meyer's work in the 1980s is credited with laying the foundation for the mathematical theory of wavelets.
- In 1984, Ingrid Daubechies made a significant contribution to the field of wavelet theory by introducing compactly supported orthogonal wavelets.
- Over the course of the late 1980s and early 1990s, Stephane Mallat's contributions to wavelet research, which include multiresolution analysis (MRA) and discrete wavelet transform (DWT), had been enhanced.
- Using wavelets for image compression, JPEG2000, which was released in 1992, highlights the usefulness of wavelets.
- In the late 1990s, wavelet-based algorithms such as the lifting scheme expanded the practical applications of wavelet theory's theoretical framework.
- At the beginning of the twenty-first century, wavelet transforms became widely utilized in a variety of fields, including biological, financial, and data analysis.
- Beginning in the middle of the 2000s, wavelets began to find applications in the field of machine learning and signal processing.
- In the 2010s, wavelet theory was being increasingly applied in fields such as computational neuroscience and cybersecurity as a result of ongoing research.
- In the years 2015–2020, wavelet-based algorithms continued to play an important role in the analysis of large data and pattern identification.
- Present day (2023): Wavelet transforms are still developing and finding applications in a wide variety of scientific and technological fields. They are helping to simplify difficult data analysis and signal processing and are becoming increasingly popular.

Over the course of this time period, the wavelet transform has developed from its beginnings as a mathematical notion to a tool that is extensively used across a variety of fields, thereby revolutionizing signal processing, data analysis, and picture compression. Its relevance in contemporary scientific and technical achievements can be attributed, in part, to its ongoing evolution and variety among its applications.

Table 1 given below provides a detailed comparison of Fourier Transform, Wigner–Ville Transform, Short-Term Fourier Transform (STFT), and Wavelet Transform characteristics concerning time-frequency analysis, localization, resolution, advantages, and limitations, allowing for comparison across these widely used signal processing techniques.

Transform	Time-frequency analysis	Localization	Resolution	Advantages	Limitations
Fourier Transform	No	Global	Fixed	Represents signals by frequency components; Useful for stationary signals	Lacks time information; Not suitable for nonstationary signals
Wigner–Ville Transform	Yes	Good	Moderate	Precise time-frequency analysis; Reveals instantaneous frequency information	Cross-term interference; Sensitivity to noise
Short-Term Fourier Transform	Yes	Moderate	Variable	Localized time-frequency information; Suitable for nonstationary signals	Fixed time-frequency resolution; Trade-off between resolution and localization
Wavelet Transform	Yes	Excellent	Variable	Localized in both time and frequency; Multiresolution analysis of signals	Offers variable resolution but computationally intensive

Table 1.
Characteristics of various transformations.

3. Types of wavelet transforms

Different types of wavelet transforms offer diverse characteristics and functionalities, tailored for specific applications in signal processing, image analysis, data compression, and more [12, 15]. Here is a list different important wavelet transforms:

Continuous Wavelet Transform: The continuous wavelet transform (CWT) is a signal analysis technique. It compares a signal to modified versions of a "mother wavelet" function. The modified versions are both scaled and translated. The CWT operates constantly across different levels and durations. This allows for the examination of signals that are not consistent and have different frequency components. The characteristics of the analysis are determined by the selection of the mother wavelet. The CWT offers accurate temporal and spectral localization. This makes it suitable for signals with sudden variations or fluctuating frequencies. However, it requires significant computational resources.

Discrete Wavelet Transform: The discrete wavelet transform (DWT) breaks down signals into different frequency components at multiple resolutions. It uses low-pass and high-pass filters to separate signals into approximate and detailed information. The DWT is useful for tasks like compression, noise reduction, feature extraction, and analyzing transitory occurrences in various domains. It provides a comprehensive analysis of different resolutions, collecting both high- and low-frequency data. The hierarchical representation of signals with varying levels of detail makes it applicable in many fields [10, 11].

Wavelet Packet Transform: The wavelet packet transform (WPT) is an expansion of the discrete wavelet transform (DWT). It offers a more thorough signal decomposition. The DWT partitions signals into approximation and detail coefficients at various scales. However, the WPT enables both high- and low-frequency sub-bands at each level of decomposition to be further decomposed. This leads to a more comprehensive examination with enhanced adaptability. It allows for a more intricate investigation of signal attributes across different frequency ranges. The WPT is utilized in several fields such as signal processing, feature extraction, and data compression. It provides improved flexibility.

Multiwavelet Transform: Multiwavelet transformations build upon wavelet transformations by using multiple functions for signal decomposition. Unlike typical wavelet transforms that use only one scaling function and its wavelet, multiwavelet transforms utilize many scaling functions and their accompanying wavelets. This approach offers advantages such as improved estimation of signals with imperfections, better symmetry, and increased flexibility in signal representation compared to single-wavelet transforms. Multiwavelet transformations find applications in various signal processing tasks, including image and signal compression, denoising, and feature extraction. Their ability to adapt and capture complex signal properties.

Curvelet Transform: The Curvelet Transform accurately represents images with complex geometric characteristics and smooth curves. It operates at multiple scales and directions, capturing delicate details in photos. It outperforms wavelet-based transforms in capturing edges, curves, and nonlinear structures. Curvelets excel at capturing curved and angular characteristics, making them suitable for portraying objects with varying orientations and scales. The curvelet transform is valuable in medical imaging, geophysical data analysis, and image processing tasks that require accurate depiction of edges and curves.

Ridgelet Transform: The ridgelet transform is a specialized transform that efficiently represents images with linear features or edges. It is better than typical wavelet-based approaches in capturing linear structures in photographs, regardless of their orientations and scales. It is particularly effective in scenarios where images consist mostly of linear features, such as in medical imaging or seismic data processing. The ridgelet transform offers a sparse depiction of linear characteristics, making it valuable in various image processing applications that require accurate identification and examination of edges or line-like formations.

Contourlet Transform: The contourlet transform is a method for representing images that captures edges and contours efficiently. It uses a multiscale and multidirectional approach. It combines wavelets with directional filter banks to create a flexible representation, particularly suitable for images with curves and edges. The contourlet transform has improved directionality and localization, making it useful in applications that require accurate depiction of edges, curves, and textures in images. It is efficient in handling complex geometric structures, making it valuable in applications like image compression, denoising, medical imaging, and other tasks that require accurate depiction of edges and textures.

Integer Wavelet Transform: The integer wavelet transform (IWT) is a variant of the discrete wavelet transform (DWT) that only works with integer data. It is useful for processing signals or images that have only integer values, like certain compression methods and embedded devices. The IWT does not require floating-point arithmetic, which makes calculations faster and more accurate since it only uses integers. This is beneficial in situations where efficiency, memory usage, or adherence to integer-based systems are important, such as in embedded devices, hardware implementations, or specific compression methods.

Wavelet transform	Use cases	Advantages	Limitations
Continuous Wavelet Transform	Nonstationary signals	Precise time-frequency localization; Analysis of signals with varying frequencies over time	Computationally intensive
Discrete Wavelet Transform	Signal compression, denoising, imaging	Efficient, hierarchical representation; Transient analysis	Loss of phase information; Boundary effects
Wavelet Packet Transform	Detailed frequency analysis	Enhanced flexibility; More detailed signal exploration	Increased computation due to higher decomposition
Multiwavelet Transform	Signal approximation and features extraction	Better symmetry; Enhanced adaptability	Increased complexity
Curvelet Transform	Medical imaging, geophysical data analysis	Superior edge and curve representation; Multiscale analysis	High computational demands
Ridgelet Transform	Seismic data analysis, edge detection	Efficient representation of linear features	Limited application in nonlinear structures
Contourlet Transform	Image compression, denoising	Excellent edge and texture representation; Flexible analysis	Computationally intensive; Increased complexity
Integer Wavelet Transform	Embedded systems, integer-based data	Faster computation, reduced memory; Integer-based processing	Limited to integer-based data; Reduced flexibility

Table 2.
Comparison of different types of wavelet transformations.

Table 2 given below highlights the distinct characteristics, applications, advantages, and limitations of each type of wavelet transform, showcasing their specific utilities across various domains.

4. Applications of wavelet transform

The relevance of the wavelet transform resides in its capacity to evaluate signals and images at many resolutions while maintaining time-frequency localization, rendering it a potent tool in several domains [14, 15]. The wavelet transform is significant and applicable in several major areas:

1. *Image Compression:* Reducing the data size of images while retaining quality by analyzing and discarding less critical image information in different frequency bands.

2. *Signal Denoising:* Removing unwanted noise from signals while preserving essential features by separating noise from the signal components at various scales.

3. *Audio Compression:* Reducing the size of audio files without significant loss of quality, vital in efficient storage and transmission of audio data.

4. *Feature Extraction in Image Processing:* Identifying and extracting meaningful features from images, such as edges or textures, for subsequent analysis or pattern recognition.

5. *Seismic Signal Analysis:* Studying seismic waves to understand subsurface structures and earthquake characteristics, aiding in geophysical exploration.

6. *Edge Detection in Image Processing:* Identifying boundaries or edges between objects in images, crucial for object recognition and image segmentation.

7. *Financial Time-Series Analysis:* Studying financial data trends, identifying patterns, and predicting market behavior for investment decisions.

8. *Speech Processing:* Analyzing speech signals for tasks like speech recognition, language translation, and voice-based interfaces.

9. *Biometric Systems:* Extracting distinctive features from biometric data (like fingerprints or irises) for identity verification.

10. *Communication Systems:* Analyzing modulated signals in communication systems for signal processing, error correction, and so forth.

11. *Pattern Recognition:* Identifying and categorizing patterns or objects in data, crucial in machine learning and computer vision.

12. *Geophysical Data Analysis:* Processing geophysical data to understand geological formations and subsurface structures.

13. *Texture Analysis in Image Processing:* Characterizing textures in images for various applications, including remote sensing and materials analysis.

14. *Nondestructive Testing:* Analyzing signals to detect flaws or defects in materials without causing damage, used in industry and materials science.

15. *Vibration Analysis:* Studying vibrations in mechanical systems for fault detection and condition monitoring in machinery.

16. *Time-Frequency Analysis in EEG Signals:* Extracting frequency information over time from EEG signals to understand brain activity patterns.

17. *Molecular Biology:* Analyzing biological signals to study genetic patterns, molecular interactions, and so on, in biological research.

18. *Fault Detection in Power Systems:* Monitoring power systems to detect and diagnose faults for maintaining grid stability.

19. *Environmental Data Analysis:* Analyzing environmental signals for studying climate patterns, ecological changes, and so forth.

20. *Video Compression:* Compressing video data efficiently for storage, streaming, and transmission.

21. *Sonar Signal Processing:* Analyzing underwater signals for navigation, target detection, and marine communication.

22. *Radar Signal Processing:* Analyzing radar signals for object detection, tracking, and navigation in aerospace and defense.

23. *Spectral Analysis:* Decomposing signals into frequency components for analyzing spectral characteristics.

24. *Image Enhancement:* Improving the quality or appearance of images for better visualization or analysis.

25. *Data Fusion:* Combining multiple sources of information to enhance data accuracy or completeness.

26. *Character Recognition:* Identifying and converting characters from images into text for OCR applications.

27. *Object Tracking:* Following the movement of objects in video sequences for surveillance or monitoring.

28. *Fractal Analysis:* Analyzing complex patterns or shapes using fractal geometry for various applications.

29. *Remote Sensing:* Using sensors to collect data from a distance for environmental or geographical analysis.

30. *System Identification:* Modeling and understanding the behavior of dynamical systems from measured data.

31. *Image Watermarking:* Embedding information into images for copyright protection or authentication.

32. *Wireless Communication Systems:* Analyzing signals in wireless networks for efficient data transmission.

33. *Image Registration:* Aligning multiple images for comparison or creating panoramic views.

34. *Anomaly Detection:* Identifying unusual patterns or events in data that deviate from expected behavior.

35. *Quality Assessment in Images:* Evaluating image quality for various applications like printing or medical imaging.

36. *Time Series Forecasting:* Predicting future values based on past data patterns in time series analysis.

37. *Motion Detection in Video:* Detecting movement in video sequences for security or activity monitoring.

38. *Hyperspectral Imaging Analysis:* Analyzing images with numerous spectral bands for detailed material identification.

39. *Structural Health Monitoring:* Monitoring structural conditions of buildings or infrastructure for maintenance.

40. *Channel Equalization:* Compensating for distortion in communication channels to recover transmitted signals.

41. *Quantum Signal Processing:* Analyzing quantum signals or information processing in quantum systems.

42. *Robotics and Vision Systems:* Processing visual data for robot guidance and control in robotics applications.

43. *ECG Signal Analysis:* Analyzing electrocardiogram signals for diagnosing heart conditions or abnormalities.

44. *Sonography Image Processing:* Enhancing and analyzing ultrasound images for medical diagnosis.

45. *DNA Sequence Analysis:* Analyzing DNA sequences for understanding genetic information and mutations.

46. *Audio Signal Separation:* Separating mixed audio sources into individual components for analysis or modification.

47. *Speaker Recognition:* Identifying individuals by analyzing characteristics of their voice patterns.

48. *Waveform Analysis:* Analyzing waveforms to understand characteristics or patterns in signals.

49. *Information Retrieval:* Extracting relevant information from large datasets or databases.

50. *Computational Neuroscience:* Applying computational methods to study brain function and neural systems.

51. *Gait Analysis:* Analyzing human walking patterns for medical, biomechanical, or forensic purposes.

52. *Gesture Recognition:* Recognizing and interpreting human gestures for human–computer interaction.

53. *Traffic Analysis and Prediction:* Analyzing traffic patterns for congestion prediction and management.

54. *Functional Magnetic Resonance Imaging (fMRI) Analysis:* Analyzing brain activity based on fMRI scans to understand brain function.

55. *Texture Synthesis:* Creating new textures based on existing ones for graphics or modeling.

56. *Sleep Pattern Analysis:* Studying sleep patterns and stages for sleep disorder diagnosis.

57. *Electroencephalography (EEG) Analysis:* Analyzing brain electrical activity for neuroscience or medical diagnostics.

58. *Antenna Array Processing*: Processing signals from antenna arrays for improved wireless communications.

59. *Intrusion Detection:* Detecting and preventing unauthorized access or attacks in computer systems.

60. *Text Mining:* Extracting useful information or patterns from large volumes of text data.

61. *Time-Frequency Analysis in Music:* Analyzing music signals to understand their frequency and time characteristics.

62. *Eye Tracking:* Tracking eye movements to understand visual attention or diagnose eye conditions.

63. *Glottal Analysis:* Studying characteristics of vocal fold vibrations for speech and voice analysis.

64. *Solar Activity Prediction:* Predicting solar activities like sunspots or flares for space weather forecasting.

65. *Image Matting:* Extracting foreground objects from an image for editing or composition.

66. *Electrocardiography (ECG) Signal Analysis:* Analyzing heart electrical activity for diagnosing cardiac conditions.

67. *Spatiotemporal Data Analysis:* Analyzing data considering both space and time dimensions for various applications.

68. *Synthetic Aperture Radar (SAR) Processing:* Analyzing radar data for high-resolution imaging in remote sensing applications.

69. *Gene Expression Analysis:* Studying patterns of gene activity to understand biological processes and diseases.

70. *Surface Defect Detection:* Identifying defects or anomalies on surfaces for quality control in manufacturing.

71. *Oceanographic Data Analysis:* Analyzing ocean data for understanding marine ecosystems, currents, and climate.

72. *Financial Volatility Analysis:* Studying fluctuations in financial markets to assess risk and volatility.

73. *ECG-based Biometric Systems:* Using ECG signals for biometric identification or authentication purposes.

74. *Structural Damage Identification:* Identifying structural damage or deterioration in buildings or infrastructure.

75. *Traffic Signal Timing Optimization:* Optimizing traffic signal timings for better traffic flow and congestion management.

76. *Human Activity Recognition:* Identifying and categorizing human activities from sensor data for various applications.

77. *Biomedical Image Fusion:* Combining multiple biomedical images for better visualization or analysis.

78. *Radio Astronomy Data Analysis:* Analyzing signals from radio telescopes for studying celestial objects or phenomena.

79. *Brain-Computer Interfaces:* Using brain signals for controlling external devices or computers.

80. *Solar Power Forecasting:* Predicting solar energy production for efficient grid management.

81. *Gesture-based Human–Computer Interaction:* Using gestures for controlling or interacting with computers or devices.

82. *Melody Extraction in Music Signals:* Extracting melodies or dominant pitches from music signals.

83. *Ionosphere Signal Processing:* Analyzing ionospheric signals for communication or navigation purposes.

84. *Neuroimaging Data Analysis:* Processing brain imaging data for studying brain structure or function.

85. *Cyber-Physical Systems Analysis:* Analyzing systems that integrate physical and computational components.

86. *Photonics Signal Processing:* Processing signals in photonics for various optical or light-based applications.

87. *Object Detection in Images:* Detecting and locating objects within images or videos for various applications.

88. *Forensic Image Analysis:* Analyzing images for forensic investigations or evidence examination.

These applications showcase the wide-ranging utility of wavelet transform across diverse fields, illustrating its pivotal role in signal processing, data analysis, and scientific research in numerous domains.

Author details

Srinivasan Ramakrishnan
Dr. Mahalingam College of Engineering and Technology, Pollachi, India

*Address all correspondence to: ram_f77@yahoo.com

References

[1] Mallat SG. A theory for multiresolution signal decomposition: The wavelet representation. IEEE Transactions on Pattern Analysis and Machine Intelligence. 1989;**11**(7):674-693

[2] Daubechies I. Ten Lectures on Wavelets. Philadelphia, PA, USA: SIAM; 1992

[3] Meyer Y. Wavelets and Operators. United Kingdom: Cambridge University Press; 1992

[4] Coifman RR, Wickerhauser MV. Entropy-based algorithms for best basis selection. IEEE Transactions on Information Theory. 1992;**38**(2):713-718

[5] Percival DB, Walden AT. Wavelet Methods for Time Series Analysis. United Kingdom: Cambridge University Press; 2000

[6] Aydemir O, Kayikcioglu T. Wavelet Transform Based Classification of Invasive Brain Computer Interface Data. Radioengineering. 2011;**20**:31-38

[7] Kharitonenko I, Xing Zhang Twelves S. A wavelet transform with point-symmetric extension at tile boundaries. In: IEEE Transactions on Image Processing. Dec 2002;**11**(12):1357-1364. DOI: 10.1109/TIP.2002.806237

[8] Nascimento EGS, de Melo TAC, Moreira DM. A transformer-based deep neural network with wavelet transform for forecasting wind speed and wind energy. Elsevier Journal of Energy. 2023;**278**:1-11

[9] Strang G, Nguyen T. Wavelets and Filter Banks. United Kingdom: Wellesley-Cambridge Press; 1996

[10] Kovacevic J, Vetterli M. Wavelets and Subband Coding. Hoboken, New Jersey, U.S: Prentice Hall; 2007

[11] Nason GP. Wavelet Methods in Statistics with R. New York, NY: Springer; 2008

[12] Grossmann A, Morlet J, Paul T. Mathematics of Wavelets for Scientists and Engineers. Boca Raton, Florida, US: CRC Press; 2012

[13] Stanković S, Falkowski BJ. Wavelets and Subband Coding. Amsterdam, Netherlands: Elsevier; 2018

[14] Zhang Y et al. Wavelet Neural Networks: With Applications in Financial Engineering, Chaos, and Classification. Boca Raton, Florida, US: CRC Press; 2020

[15] Antonini M et al. Wavelet Analysis and its Applications. Boca Raton, Florida, US: Springer; 2022

Chapter 2

Improvement of the Global Positioning Accuracy with Miniaturized Strap-Down INS Systems through Wavelet Filtering of Data from MEMS Inertial Sensors

Bianca-Gabriela Antofie, Marius-Ciprian Larco and Teodor Lucian Grigorie

Abstract

In this chapter, the correlation between the issue of perturbed data acquired from miniaturized inertial sensors and the wavelet filtering technique is investigated. The market growth of micro-electro-mechanical systems (MEMS) has impacted various fields, and its potential application in strap-down inertial navigation systems (INS) could not be overlooked. Despite the apparent benefits of dimension and price reduction, the utilization of miniaturized inertial sensors for manufacturing the inertial measurement unit (IMU) entails certain drawbacks: the output signals are usually corrupted with different types of errors, which distort the real navigation information. The proposed case focuses on the suitability of wavelets for denoising the perturbed IMU signals before being erroneously processed by the navigation algorithm. The applicative part consisted in implementing sensor software models in Simulink and testing various wavelet filters. Furthermore, to fully assess the efficacy of the wavelet denoising technique, the model of a SDINS employing a MEMS-based IMU was established in Simulink. The evaluation involved comparing the attitude, position and speed components obtained before and after the denoising procedure with those of an ideal model linked to constant inputs. The results demonstrated the effectiveness of the proposed association in terms of positioning accuracy, signal characteristics improvement and computation complexity.

Keywords: strap-down INS, MEMS-based inertial measurement unit, wavelet filtering of MEMS data, global positioning, positioning accuracy improvement

1. Introduction

Micro-electro-mechanical systems gained immensely in popularity within sensor manufacturing technologies, primarily owing to their distinctive properties arising from the integration of both mechanical and electrical devices on a single chip. In the same vein, the strap-down architecture of inertial navigation systems has diminished the mechanical complexity of previous platform versions, consequently driving the demand for smaller inertial sensors [1]. The adoption of MEMS technology for the IMU fabrication of strap-down inertial navigation systems is primarily generated by the significant advantages it offers, including reduced size, cost-effectiveness, scalability in serial production, and low power consumption. However, the miniaturization process of electronic devices leads to enhanced sensitivity to external environmental factors such as temperature, humidity, and vibrations. Consequently, this increased sensitivity leads to the propagation of errors and noises in the output acceleration and angular rate signals [2]. Instead of a continuous and clear evolution representation of the measured dataset, the useful signal is often submerged in the noise. Conventional filtering methods usually prove to be inefficient and may even cause signal distortion due to the tendency of noise frequency bands to overlap with the frequency band in which the dynamics of the monitored vehicle are located [3].

In the past decade, the field of signal processing has directed its attention towards the relatively modern discovery of wavelet theory. The wavelet transform, based on multiresolution analysis, decomposes a signal into wavelet components at various scales, enabling the representation of different levels of details within the signal. Numerous applications have been found for wavelets in fields such as image processing, denoising, feature extraction, and compression, enabling efficient analysis of oscillatory phenomena in various data types [4]. Because of their ability to effectively denoise signals while preserving critical information, wavelets present a promising option to address the rigorous requirements of inertial navigation. Hence, the investigation of wavelet filters' behavior regarding perturbed inertial data is a worthwhile subject for research.

The main objective of the experimental part was to evaluate the effectiveness of pre-filtrating the outputs of inertial sensors before incorporating them as inputs into the strap-down navigation algorithm, with a focus on the essential improvement in positioning accuracy. To set the context for the practical part, this study first provides a comprehensive overview of the types of errors associated with MEMS inertial sensors. Subsequently, a mathematical foundation is established to support the application of the wavelet denoising technique.

The experimental part presents how two MEMS sensor models—an accelerometer and a gyroscope—were software implemented and simulated with constant inputs using MATLAB/Simulink. The resulting perturbed output signals underwent decomposition using various wavelet functions, and their performances were assessed through the calculation of specific estimator measures. In the second part, the strap-down navigation algorithm was fully developed in the Simulink software environment and tested in association with the wavelet filter. The outcomes of the simulations furnished a comprehensive insight into the implications of MEMS architecture on the output data precision of the strap-down algorithm. Moreover, the extent to which wavelet techniques succeed in enhancing the overall accuracy was also rigorously assessed.

As important limitations of the proposed work can be mentioned the partial removal of the noise perturbing the inertial sensors and the offline tuning of the wavelet transform, which means that the noise coming from various sources operating near the sensors during flight cannot be limited through a real-time tuning of the involved wavelet mechanisms.

2. Preliminary analysis of the miniaturization issues

As MEMS-based sensors are scaled down to smaller dimensions, they become inherently more sensitive to external environmental factors, which can lead to adverse effects on their performance. One of the primary concerns arising from miniaturization is the increased susceptibility to external noise sources. Due to their reduced size, MEMS inertial sensors exhibit higher sensitivity to temperature fluctuations, humidity variations, and mechanical vibrations. These environmental influences can introduce unwanted perturbations and inaccuracies in the sensor's output signals, compromising the overall measurement accuracy. Moreover, as the size of MEMS inertial sensors decreases, there is a consequent reduction in their mass and inertial properties. This reduction may lead to diminished signal-to-noise ratios, making it challenging to distinguish the desired signal from noise effectively. Consequently, the signal of interest may become masked by the noise, hindering the accurate extraction of relevant information [5].

In general, errors in MEMS inertial sensors can be categorized into two main groups: deterministic errors and stochastic errors. Deterministic errors are systematic and predictable errors that result from inherent characteristics of the sensor or external influences. They include bias error, scale factor error, misalignment error, and cross-coupling error. Stochastic errors, on the other hand, are random and unpredictable variations in the sensor's output caused by environmental factors and noise sources. Some examples of stochastic errors are random noise, drift, vibration noise, and quantization noise [6].

Understanding and characterizing these errors is crucial for accurate sensor calibration and data interpretation. Their integration with the original signal introduces inaccuracies and uncertainties in the measured data, directly impacting the overall performance of the strap-down navigation system and potentially compromising the success of its mission. **Figure 1** visually illustrates the errors that overlap a MEMS gyro's original signal, resulting in a corrupted output. This graphical representation emphasizes the importance of utilizing denoising techniques, such as wavelet filtering, which will be widely discussed during the next section.

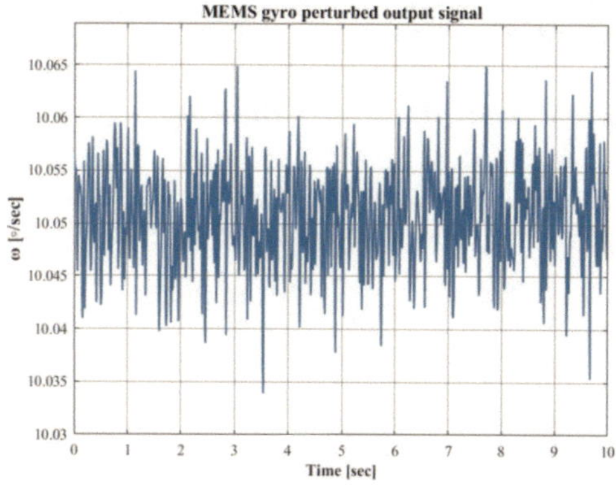

Figure 1.
Corrupted output signal of a MEMS gyro.

3. Wavelet denoising method

3.1 Mathematical background of wavelets

A wavelet is a mathematical function characterized by its oscillating pattern, where the amplitude initiates from zero, undergoes multiple cycles of increase and decrease, and eventually returns to zero. It allows signals to be analyzed and transformed into different frequency components, providing a way to capture both high and low-frequency oscillations in a signal.

Wavelets are particularly advantageous when dealing with signals that exhibit non-stationary or localized oscillatory behavior [4]. **Figure 2** presents a visual representation of the most common types of wavelet functions.

The wavelet transform is a powerful mathematical tool used in signal processing to decompose a signal into its frequency components. There are two types of wavelet transforms: continuous wavelet transform (CWT) and discrete wavelet transform (DWT). The CWT allows for continuous variation in the scale of the wavelet function, providing a detailed representation of the signal's frequency content at different time points. On the other hand, the DWT breaks down the signal into discrete scales, which is computationally more efficient and suitable for practical applications [7].

For denoising the output signals of our miniaturized inertial sensors, the DWT is particularly well-suited. The DWT can effectively capture localized and non-stationary oscillations, making it advantageous for processing signals affected by external environmental factors and noise. By decomposing the signal into various scales, the DWT allows for a more precise analysis and separation of noise from the useful signal, resulting in improved denoising performance [8].

In the field of signal processing, various transforms, such as the Fourier Transform (FT) and the Fast Fourier Transform (FFT), have been widely employed for frequency domain analysis. While these transforms are effective in revealing the frequency content of a signal, they possess certain limitations when dealing with signals containing localized or non-stationary features [9]. Unlike the DWT, which decomposes signals into different frequency components with varying time resolutions, the FT and FFT provide a fixed frequency resolution across the entire signal, as can be observed in **Figure 3**.

This characteristic restricts their ability to effectively capture localized features and may lead to signal distortion when dealing with non-stationary signals. The DWT's adaptability to a wide range of signal characteristics, along with its ability to efficiently handle real-world data, makes it a superior choice for various signal processing tasks, including denoising, feature extraction, and pattern recognition in practical applications.

3.2 Systematic implementation of wavelet filtering

The process of filtering a perturbed signal with a wavelet function begins with selecting a suitable type of function. The choice should be made considering the

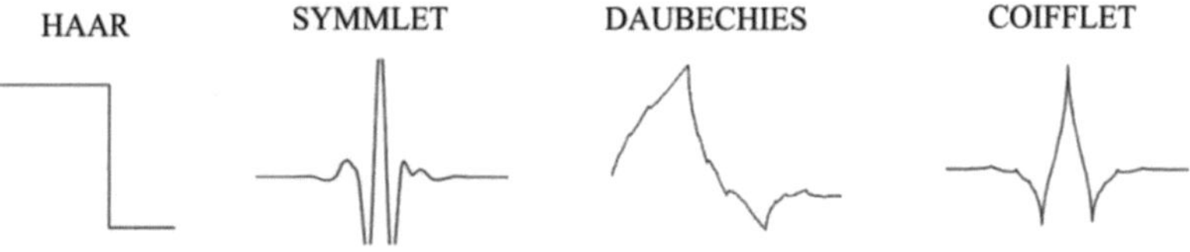

Figure 2.
Examples of wavelet functions.

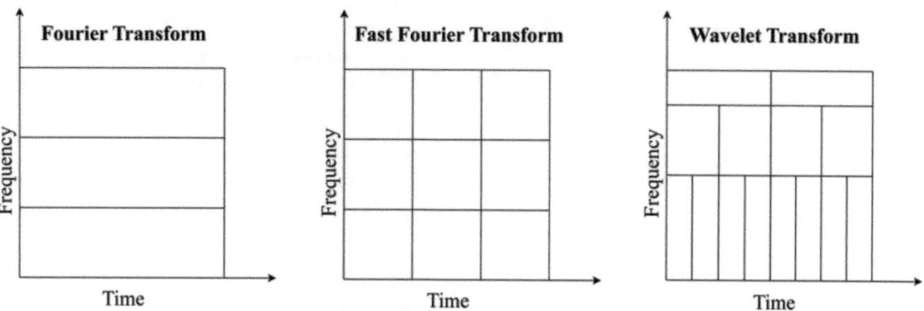

Figure 3.
Varieties of time-frequency analysis techniques.

specific characteristics of the noise and signal, as well as the desired denoising outcome; however, it is advisable to test multiple wavelets as they may exhibit distinct behaviors depending on the specific case.

When the chosen discrete wavelet transform (DWT) is applied to the signal, it operates as a sequence of high-pass and low-pass filters, leading to the decomposition of the signal into several sub-bands, each representing distinct frequency ranges. This procedure yields wavelet coefficients, categorized as either approximation (cA) responsible for capturing low-frequency information or details (cD) responsible for capturing high-frequency information (**Figure 4**) [9].

The process of decomposition can be iteratively performed by applying the discrete wavelet transform (DWT) to the obtained approximation coefficients (cA) from the initial decomposition stage. This recursive application results in a multi-level wavelet decomposition, as illustrated in **Figure 5**. Each level of the decomposition provides a new set of approximation coefficients, capturing lower frequency components successively. This hierarchical approach facilitates the analysis of the signal at various scales and resolutions, offering a comprehensive insight into its frequency content.

During this stage, the concept of thresholding becomes crucial. Once the decomposition process is complete, a specific threshold is established, and each absolute

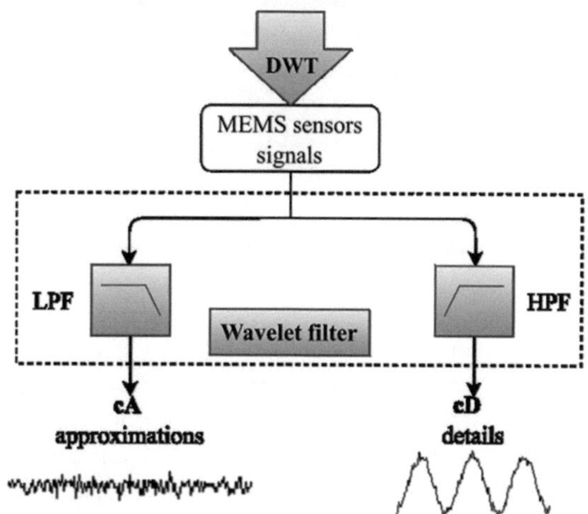

Figure 4.
DWT signal decomposition.

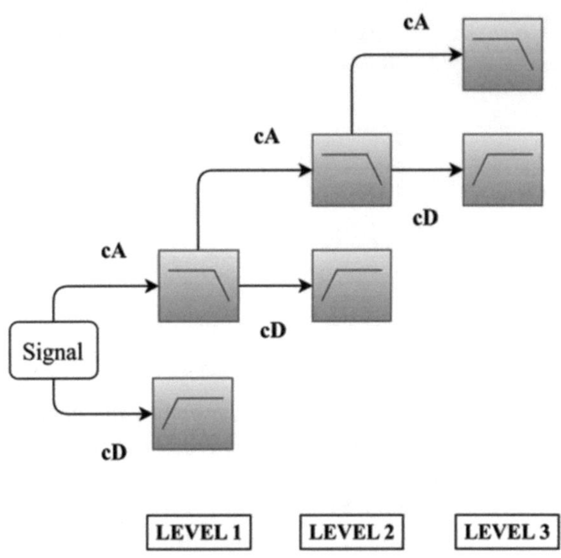

Figure 5.
Iterative process of wavelet decomposition.

value of the wavelet coefficient is subjected to comparison with this threshold. The threshold value can be computed using the following formula [10]:

$$\lambda = \sqrt{2logn}\sigma \tag{1}$$

where: n – number of samples, σ – standard deviation of the signal.

In wavelet denoising, two common thresholding techniques are used: soft thresholding and hard thresholding. Soft thresholding shrinks wavelet coefficients below the threshold to zero, while hard thresholding sets coefficients below the threshold to zero. The choice between the two depends on the signal characteristics and denoising goals. In the current case, soft thresholding is preferred as it offers a more gradual suppression of noise compared to hard thresholding. The perturbed navigation signals often contain various levels of noise, and the abrupt removal of coefficients through hard thresholding may cause a significant loss of useful signal information. This choice was influenced by previous successful implementations, notably in Refs. [11, 12], which also employed soft thresholding. Eq. (2) shows the soft thresholding function [10]:

$$T_{\lambda}^{soft} = \begin{cases} (u - sign(u))\lambda if|u| \geq \lambda \\ 0othrewise \end{cases} \tag{2}$$

After the thresholding phase in the wavelet filtering procedure, the signal is restored from the adjusted wavelet coefficients. By performing the inverse wavelet transform to these coefficients, the original signal is reconstructed with reduced noise and preserved essential characteristics. This updated signal offers an improved representation of the underlying useful information, making it suitable for further analysis and applications, such as navigation and control systems.

The wavelet denoising process outlined above applied to our research case can be summarized into five key stages, as illustrated by **Figure 6**.

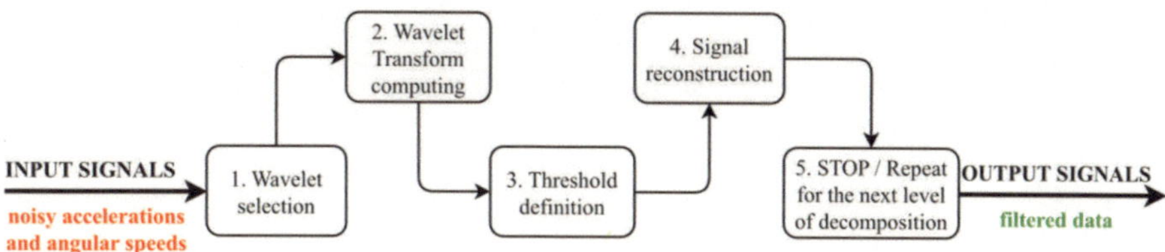

Figure 6.
Wavelet denoising method.

3.3 Filter performance evaluation criteria

In the denoising process, a crucial concern arises regarding the suitability of the chosen wavelet function and the optimal level of decomposition to repeat the procedure. To address this, several evaluation metrics can be employed, offering relevant insights into the filter's performance in signal reconstruction. Throughout this research paper, the selected evaluation criteria include the mean square error (MSE) and signal-to-noise ratio (SNR), calculated with formulas (3) and (4) [13]:

$$MSE = \frac{1}{n}\sum_{i=1}^{n}\left[x(i) - \tilde{x}(i)\right]^2, \tag{3}$$

$$SNR = 10 \cdot log_{10}\left\{\frac{\sum_{i=1}^{n}(x(i))}{\sum_{i=1}^{n}\left[x(i) - \tilde{x}(i)\right]^2}\right\}, \tag{4}$$

where: n – number of samples, $x(i)$ – ideal, unfiltered signal, $\tilde{x}(i)$– filtered signal.

Reference [11] involved comparing SNR values obtained from different wavelet functions applied to noisy acceleration signals, facilitating the ranking of their performance. Also, the addition of the last point deviation from the reference distance confirmed the previous results. In our scenario, the MSE and SNR parameters assist us in identifying the optimal decomposition level by selecting values that adhere to specified limits, while also enabling a performance comparison between the two wavelet functions (**Figure 7**).

The wavelet filter's performance, in terms of positioning accuracy improvement, can be assessed by directly analyzing the navigation information before and after the denoising procedure. This evaluation included a comparison with an ideal model of the strap-down navigation algorithm that utilizes unperturbed inputs. The ideal model gave conceptual values representing the expected outputs if the sensor data were not affected by perturbations. By contrasting the denoised navigation information with these conceptual values, the effectiveness of the wavelet filter in enhancing positioning accuracy can be determined.

4. Software configuration of the MEMS-based IMU and wavelet filter association

4.1 Inertial sensors software implementation

An IMU is the main sensing component of any INS and it is generally composed of two triads of inertial sensors—accelerometers for sensing the specific force and gyros

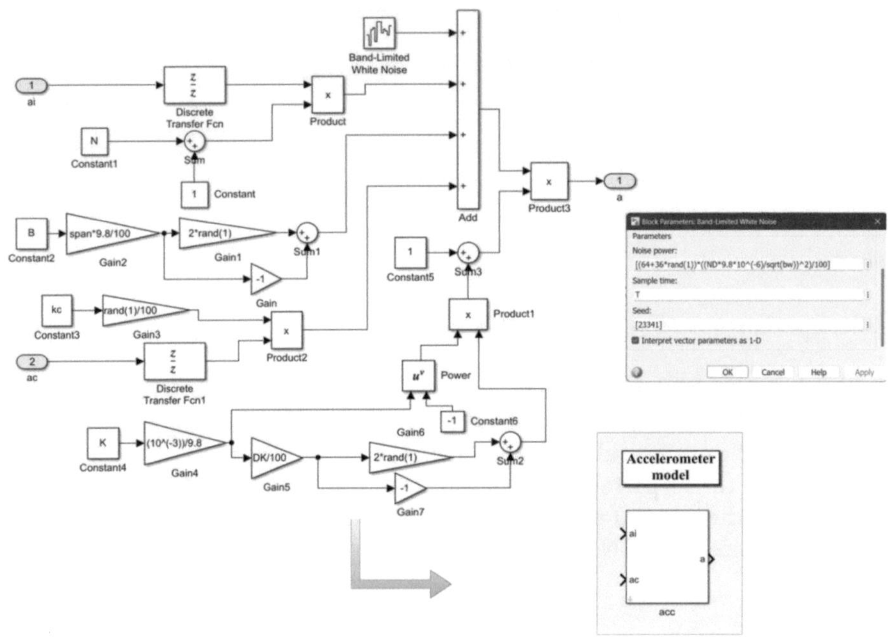

Figure 7.
MATLAB/Simulink accelerometer model.

for the angular rate, both being rigidly attached to the body of the vehicle. In order to achieve precise and realistic results in a comprehensive study that mirrors the actual conditions of the navigation system, including the inherent errors of the sensors employed, it becomes imperative to establish an equivalent model. To create the software representation of an IMU, the process was initiated by implementing an accelerometer and a gyro in MATLAB/Simulink using IEEE standardized models [3, 14].

The accelerometer model follows the equation:

$$a = (a_i + N \cdot a_i + B + k_c a_c + \nu)\left(1 + \frac{\Delta K}{K}\right), \tag{5}$$

where: a – output acceleration (perturbed signal), N – misalignment of the sensitive axis, a_i – acceleration applied along the sensitive axis, B – bias, k_c – cross-axis sensitivity, a_c – cross-axis acceleration, B – bias, ν – noise, K – scale factor, ΔK – calibration error of the scale factor.

The gyro model follows the following formula:

$$\omega = (\omega_i + S \cdot a_r + B + \nu)\left(1 + \frac{\Delta K}{K}\right) \tag{6}$$

where: ω – output angular speed, ω_i – input angular speed, S – sensitivity to the acceleration a_r applied upon an arbitrary direction, a_r – resultant of the accelerations applied on the three directions of the accelerometric triad of the strap-down inertial system, B – bias, ν – noise, K – scale factor, ΔK – calibration error of the scale factor (**Figure 8**).

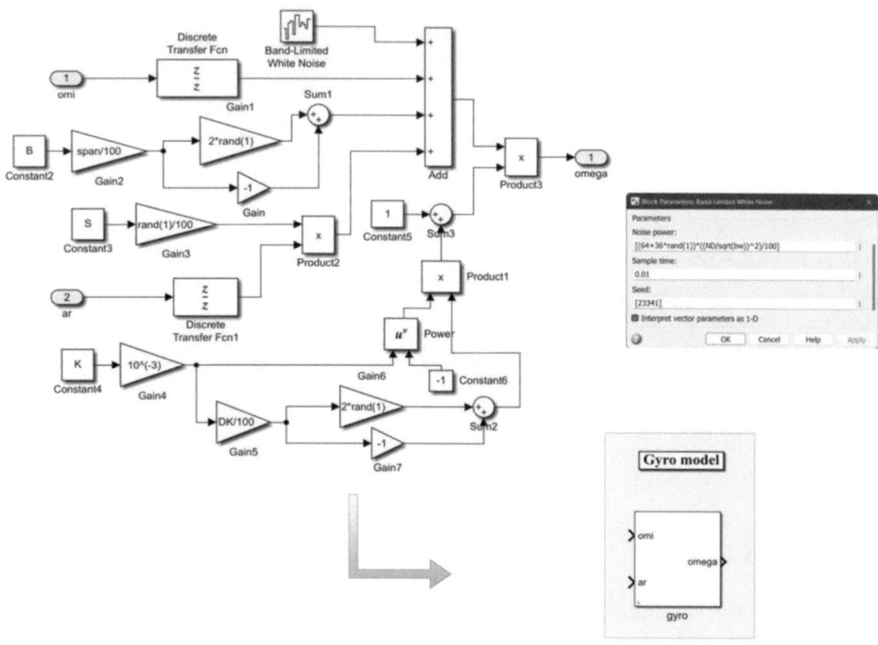

Figure 8.
MATLAB/Simulink gyro model.

In the upcoming part, two sensor models popular on the MEMS market were implemented: the S1221x-025 accelerometer and the ADXRS150 gyro. These selections were made to ensure the simulation's accuracy and relevance, as these sensors represent the prevailing MEMS technology and are commonly employed in various aerospace applications. **Tables 1** and **2** display the technical specifications of the chosen MEMS models, as well as the corresponding variable names used throughout the software configuration [15, 16].

To assemble the inertial measurement unit, three sensor components from each category were incorporated and interconnected into two triads, as depicted in

S1221x-025 SPECIFICATIONS	
Measurement range (span) [g]	±25
Scale factor (K) [mV/g]	160
Bandwidth (bw) [Hz]	1500
Noise density (ND) $[\mu g/\sqrt{Hg}]$	25
Bias (B) [% from span]	2
Scale factor error (DK) [% from scale factor]	2
Cross-axis sensitivity (S) [%]	3
Misalignment of the sensitive axis [rad]	0
Power (P) [mW]	50

Table 1.
Accelerometer data sheet parameters.

ADXRS150 SPECIFICATIONS	
Measurement range (span) [°/s]	±150
Scale factor (K) [mV/(°/s)]	12.5
Bandwidth (bw) [Hz]	80
Noise density (ND) $\left[(°/\mathbf{s})/\sqrt{\mathbf{Hg}}\right]$	0.05
Bias (B) [% from span]	1.4
Scale factor error (DK) [% from scale factor]	0.7
Cross-axis sensitivity (S) [(°/s)/g]	0.2
Power (P) [mW]	80

Table 2.
Gyro data sheet parameters.

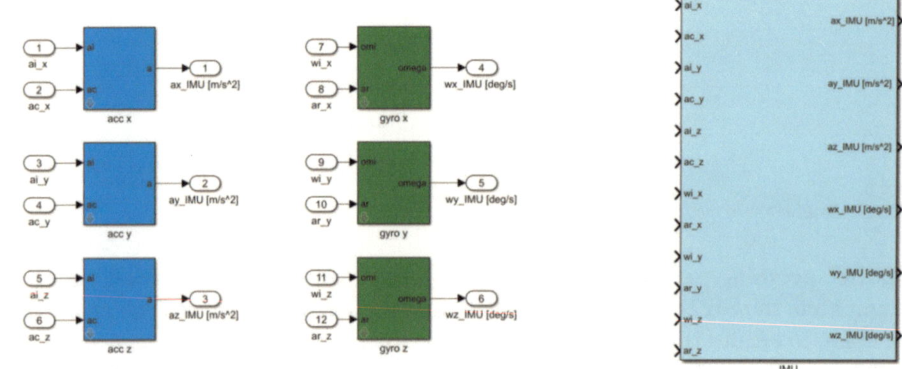

Figure 9.
MATLAB/Simulink IMU model.

Figure 9. This arrangement was streamlined into a subsystem labeled as 'IMU' for ease of representation and analysis.

4.2 Initial evaluation of the wavelet filter

In the preliminary investigation of the correlation between the wavelet denoising technique and the output signals generated by the MEMS sensors, the following constant inputs were applied to the accelerometer and gyro models for a simulation time of 10 seconds: 5 m/s^2, respectively 0.2°/s. The a_c and a_r parameters were set to zero in order to ease the calculus. **Figure 10** depicts the resulting perturbed signals. As anticipated, the mean values of both signals do not match the intended values, attributed to the specific errors of MEMS sensors.

The aim of this section is to test the individual performance and the corresponding optimal level of decomposition of two wavelet functions: Daubechies and Symmlet. These wavelets, known for their established orthogonality, extensive research, and versatility, emerge as the primary candidates for initial experimentation. This will be accomplished using MATLAB/wavelet analyzer, a specialized software that offers comprehensive tools for conducting in-depth wavelet analysis. Initially, the

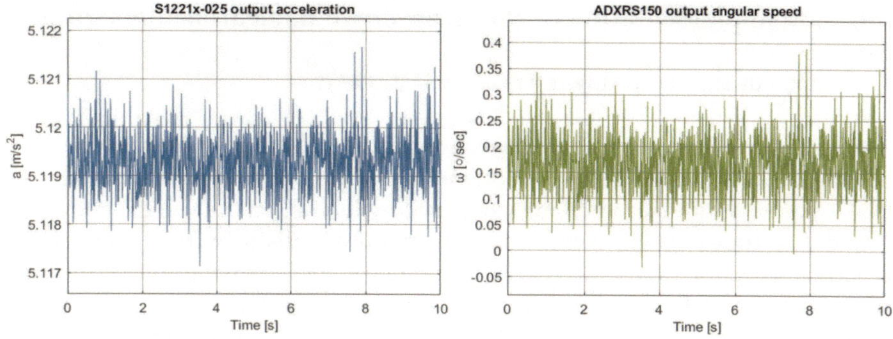

Figure 10.
Simulated perturbed outputs of the MEMS sensors.

Wavelet decomposition level	Daubechies		Symmlet		Daubechies		Symmlet	
	MSE ($\cdot 10^{-3}$)	SNR (dB)	MSE ($\cdot 10^{-3}$)	SNR (dB)	MSE ($\cdot 10^{-4}$)	SNR (dB)	MSE ($\cdot 10^{-4}$)	SNR (dB)
1	14.23	23.95	19.45	23.24	25.80	20.56	26.17	20.05
2	14.16	24.43	23.31	24.12	17.51	21.03	17.20	20.76
3	14.10	24.88	22.58	24.87	13.61	21.45	13.31	21.35
4	14.02	25.01	19.17	25.02	11.56	21.77	10.99	21.86
5	13.87	25.40	**14.23**	**25.56**	10.90	21.90	**10.95**	**21.87**
6	**13.55**	**25.45**	18.77	25.57	**10.18**	**22.18**	11.24	21.40
7	13.69	25.44	22.05	25.44	10.77	22.10	12.11	21.54

Table 3.
Wavelet filtering performance indicators.

acceleration signal was imported into the wavelet analyzer, and the Daubechies function, along with the soft thresholding option and a seven-level decomposition, were chosen. The process was replicated for the Symmlet function. Similarly, identical steps were undertaken for the angular speed signal. **Table 3** presents the values for MSE and SNR parameters calculated along each of the decomposition processes.

In general, a higher SNR value (over 20 dB) indicates a stronger signal relative to the noise, making the signal more distinguishable from the noise and resulting in improved performance for navigation data processing applications [17]. On the contrary, reducing the MSE implies achieving better results. Usually, the optimal level of decomposition is found depending on the lowest MSE and highest SNR values. Based on the data presented in the table, it is evident that the optimal level of decomposition for the Daubechies wavelet was determined to be 5, while for the Symmlet wavelet, it was found to be 6. The comparison of the results obtained by the two functions for each sensor case indicates the superior performance of the Daubechies function. As a result, the Daubechies function will be used in the denoising procedure of the entire navigation algorithm.

Significantly, it is worth noting that in [11, 12], Daubechies wavelets demonstrated remarkable success in denoising comparable signals. In [11], the SNR values were around 13 dB, suggesting a signal that was more susceptible to noise contamination and weaker in comparison to our signals.

5. Software implementation of the strap-down navigation algorithm

5.1 Definition of the strap-down problem statement

The strap-down architecture presents a navigation challenge centered on precisely estimating the position, velocity, and orientation of the monitored vehicle. This estimation is based on the measurements acquired by its IMU in the vehicle body frame (SV) and then transforming them relative to an initial reference frame, known as the local vertical geodetic frame (SG). The main challenge in this paper is to eliminate the errors introduced by these sensors in the navigation algorithm.

The algorithm can be divided into two main parts, each with its corresponding inputs. The first part involves attitude determination, which utilizes the angular speed components provided by the triad of gyros. The second part is dedicated to global position and velocity estimation, relying on the acceleration components returned by the triad of accelerometers.

5.2 Attitude determination algorithm

The algorithm is intended to use the angular rate measurements obtained from the gyros. By solving a differential equation, the algorithm calculates and provides the attitude angles of the monitored vehicle as outputs. For the current case, the Wilcox method using a matrix representation was chosen to solve the Poisson equation for attitude determination [18].

$$\dot{R}_v^l = R_v^l \cdot \begin{bmatrix} 0 & -\omega_z & \omega_y \\ \omega_z & 0 & -\omega_x \\ -\omega_y & \omega_x & 0 \end{bmatrix}, \tag{7}$$

where:

R_v^l – direction cosine matrix which assures the shift between the vehicle and the local horizontal frame.

$\omega_x, \omega_y, \omega_z$ – the components of the angular speed provided by the strap-down gyrometers triad.

Eq. (7) states the recursive calculation of the R_v^l matrix t_n iteration step considering the t_{n-1} elements of the matrix. Hence, the Wilcox method can be described as an iterative integration technique, resulting in the following form of the previous relation:

$$R_v^l(t_n) = R_v^l(t_{n-1}) \cdot T(t_n) \tag{8}$$

where $T(t_n)$ is a transition matrix defined in [19]:

$$T(t_n) = \begin{bmatrix} 1 & -\Delta\phi_z(t_n) & \Delta\phi_y(t_n) \\ \Delta\phi_z(t_n) & 1 & -\Delta\phi_x(t_n) \\ -\Delta\phi_y(t_n) & \Delta\phi_x(t_n) & 1 \end{bmatrix} \tag{9}$$

Notations: $\Delta\varphi x$, $\Delta\varphi y$, $\Delta\varphi z$ – the angular increments from the roll, pitch and yaw axis calculated as follows [18]:

$$\begin{cases} \Delta\phi_x(t_n) = \int_{t_{n-1}}^{t_n} \omega_x dt = \omega_x(t_n)\cdot(t_n - t_{n-1}) = \omega_x(t_n)\cdot\Delta t \\ \Delta\phi_y(t_n) = \int_{t_{n-1}}^{t_n} \omega_y dt = \omega_y(t_n)\cdot(t_n - t_{n-1}) = \omega_y(t_n)\cdot\Delta t \\ \Delta\phi_z(t_n) = \int_{t_{n-1}}^{t_n} \omega_z dt = \omega_z(t_n)\cdot(t_n - t_{n-1}) = \omega_z(t_n)\cdot\Delta t \end{cases} \quad (10)$$

where $\Delta t = t_n - t_{n-1}$ is the calculation step used in numerical integration.

Besides the numerical integration of the Poisson equation, the attitude algorithm requires an orthonormalization procedure for the R_v^l matrix, which needs to be updated at each iteration step by imposing the following conditions:

$$R_v^l \cdot \left(R_v^l\right)^T = \left(R_v^l\right)^T \cdot R_v^l = I_3 \quad (11)$$

Conceptually, this condition is valid if the truncation order of the integration method goes to the infinite (m $\rightarrow \infty$). Nonetheless, the current case uses a finite value for m (m = 1) a truncation error arises, resulting in an approximate version of the rotation matrix, $\hat{R}_l^v$. For this reason, at each iteration of the algorithm, an approximate matrix X of $\hat{R}_l^v$ has to be found to avoid the transition of the rectangular trihedral into a quasi-rectangular one.

According to Ref. [18, 20], the expression of X is given by:

$$X = \hat{R}_v^l \cdot \left[\left(\hat{R}_v^l\right)^T \cdot \hat{R}_v^l\right]^{\frac{-1}{2}} \quad (12)$$

By defining the error matrix [16]:

$$E = \left(\hat{R}_v^l\right)^T \cdot \hat{R}_v^l - I_3 \quad (13)$$

The final expression of X becomes [19]:

$$X = \hat{R}_v^l \cdot \left(I_3 - \frac{1}{2}E + \frac{3}{8}E^2 - \frac{5}{16}E^3 + \frac{25}{128}E^4 - \frac{63}{256}E^5 \ldots\right) \quad (14)$$

The rotation matrix of the local horizontal frame→vehicle referential transformation R_l^v can be obtained by transposing the X matrix. Moreover, the vehicle attitude angles can be calculated using the elements of R_l^v $(r_{ij})_{1\le i,j\le 3}$ and the following relations [17]:

$$\varphi = arctg\left(\frac{r_{23}}{r_{33}}\right), \theta = arcsin\,(-r_{13}), \psi = arctg\left(\frac{r_{12}}{r_{11}}\right) \quad (15)$$

The subsequent stage involved the software implementation of the Wilcox method in MATLAB/Simulink using the S-function block "WilcoxM.m" [18]. The entire scheme was simplified within the "Attitude" subsystem, which handles the attitude determination aspect of the navigation problem. The block processes the angular speed components into the Euler angles and the corresponding rotation matrix (**Figure 11**).

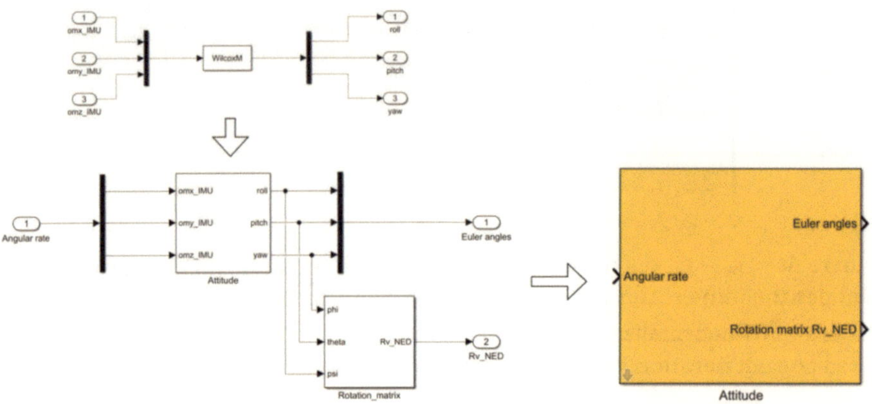

Figure 11.
Wilcox method software implementation.

5.3 Position and velocity estimation algorithm

The second part of the algorithm involves utilizing the linear accelerations along the three axes to compute the speed and position components of the monitored vehicle through integration. Because of the strap-down architecture, the acceleration components are sensed within the vehicle body frame and must be converted into the local North-East-Down frame (NED). This transformation is achieved by multiplying the acceleration vector with the previously determined rotational matrix (**Figure 12**).

However, an inconvenience arises with accelerometers as they cannot distinguish between the kinematic acceleration $\vec{a}_c$ and the gravitational acceleration $\vec{g}$, instead, they measure the resultant of both, given by $\vec{a} = \vec{a}_c + \vec{g}$. To address this issue, a model for the Earth's gravitational field is established to correct the accelerations and separate the kinematic and gravitational components. According to [18, 21], the components of gravitational acceleration in the NED frame are given by:

$$\begin{aligned} g_x &\approx 0 \\ g_y &\approx 0 \\ g_z &\approx 9.7803 + 0.0519 \cdot sin^2\phi - 3.086 \cdot 10^{-6} \cdot h, \end{aligned} \tag{16}$$

where: ϕ – geodetic latitude, h – vehicle altitude.

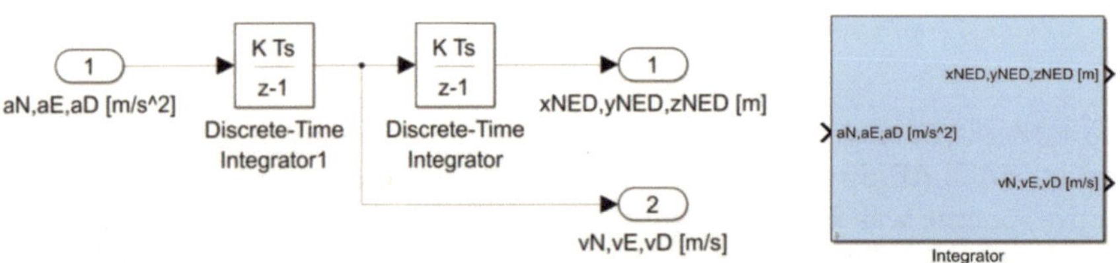

Figure 12.
'Integrator' block.

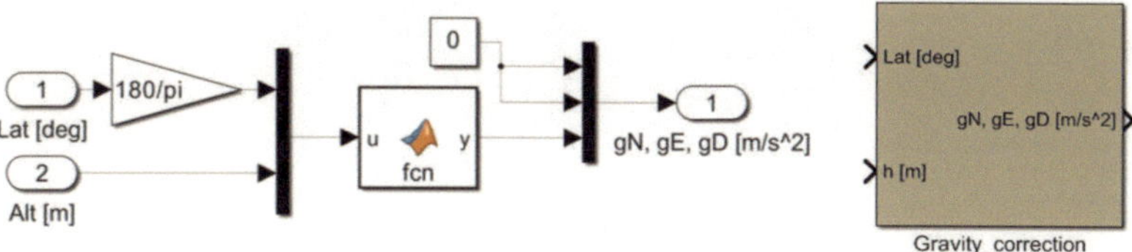

Figure 13.
'Gravity_correction' block.

Following the correction procedure, the accelerations can now be integrated to compute the speed and position in the local horizontal frame (**Figure 13**). To determine the global position of the vehicle, including its latitude, longitude, and altitude, the position coordinates in the NED frame must be converted to the SG frame. This conversion involves a two-step process: first, from NED to the Earth-Centered-Earth-Fixed (ECEF) frame, and then from ECEF to SG. For these transformations, iterative algorithms defined in [22] were employed.

The software implementation of the position and velocity estimation algorithm follows a systematic approach. Each step mentioned above will be represented by a specific block, which reveals the underlying mathematical relationships and includes a user interface for inputting initial data (**Figure 14**).

The integration block utilizes the initial values of speeds in the NED frame, the initial altitude, and the integration step [23].

The correction block requires latitude and altitude information obtained during the preceding iteration and applies the corrective model at each step (**Figure 15**) [22].

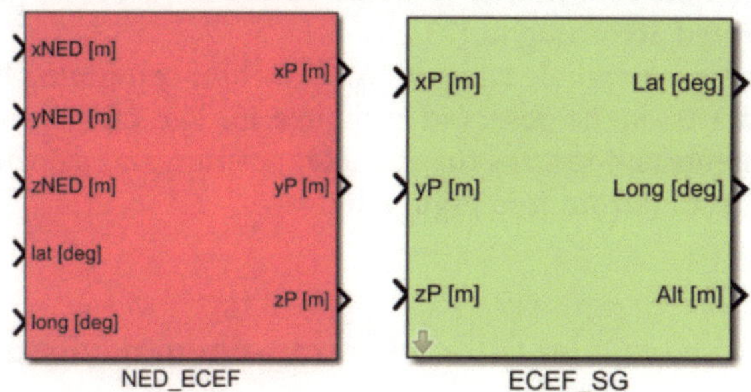

Figure 14.
'NED_ECEF' and 'ECEF_SG' conversion blocks.

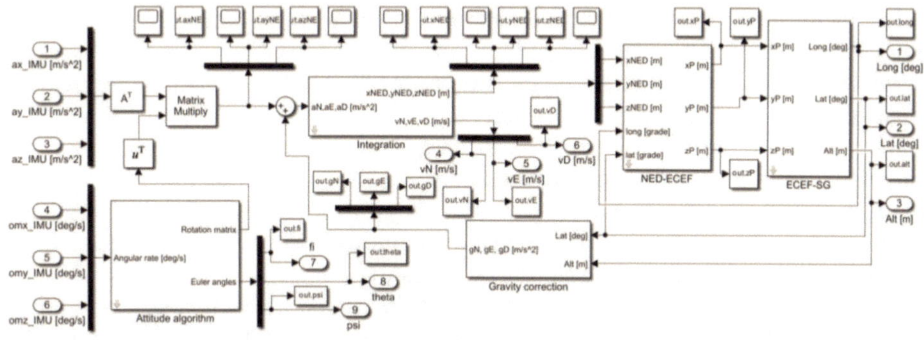

Figure 15.
MATLAB/Simulink scheme of the strap-down navigation algorithm.

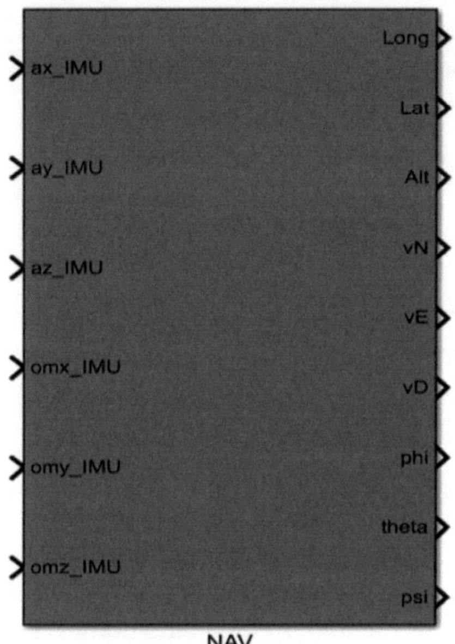

Figure 16.
'NAV' simplified block of the navigation algorithm.

Additionally, the NED → ECEF and ECEF→SG conversion blocks utilize for the current step the geodetic coordinates obtained before for accurate processing. They were also implemented according to [23].

The final scheme of the whole navigation algorithm, combining both attitude and global positioning parts can be observed in **Figure 15**. The subsystem 'NAV' that uses the IMU measurements and returns the attitude, position and velocity of the monitored vehicle was created from it in **Figure 16**.

6. Validation of the wavelet filter and strap-down navigation algorithm integration

6.1 Flight scenario definition

The focus of this chapter is on the strap-down navigation system designed for deployment on aerial vehicles engaged in short-duration missions. Consequently, in the software-based experimental phase, the emphasis is placed on simulating real-life scenarios. Reference [12] serves as compelling evidence that the investigation of digital methods to optimize inertial sensor data is most effectively conducted by applying them in practical scenarios and analyzing their impact on the ultimate attitude and position results. To ensure this, specific constant acceleration and angular rate inputs have been selected for our study to mimic the conditions experienced during take-off and level-flight ascent: $f_x = 5\ m/s^2$, $f_y = 0\ m/s^2$, $f_z = -9.82\ m/s^2$, $om_x = om_z = 0^\circ/s$, $om_y = 0.2^\circ/s$. The algorithm was also initalized with the following variables: $v_N = 4\ m/s$, $v_E = 0\ m/s$, $v_D = -2\ m/s$, $\varphi_0 = \theta_0 = \psi_0 = 0$. The starting point of the vehicle was set to 25.457319° longitude, 44.926899° latitude, and 20 m altitude. The duration of the mission (or the simulation time) was established to 60 seconds.

Observations:

- the z-axis specific force value was determined using the initial geodetic parameters, employing the relationship (16);
- initial attitude angles equal to 0 imply that the body-fixed frame is aligned with the NED frame;
- according to the initial NED speed components, the output latitude should encounter a slight rise, the longitude should remain close to zero and the altitude should significantly increase.

6.2 Generation of IMU perturbed data

To generate the perturbed MEMS outputs, referred to as "real" data, the aforementioned constant inputs were fed into the Simulink "IMU" block through the two MEMS models earlier mentioned and configured. The resulting signals from the simulation have been depicted in **Figure 17**.

The wavelet function chosen for denoising was Daubechies due to the previous results, and a 6-level decomposition was applied, resulting in the filtered signals shown in **Figure 18**.

6.3 Simulation detection cases

The proper evaluation of the filter's effectiveness involved conducting a numerical simulation in three distinct cases based on the type of inputs used for the algorithm in Simulink:

- **Ideal Case:** The 'NAV' block uses constant sensor inputs, representing the numerical values defined in the first section.
- **Real/Perturbed Case:** The 'NAV' block employs the perturbed inputs defined in **Figure 17**, applying numerical values at the entrance of the sensor models.

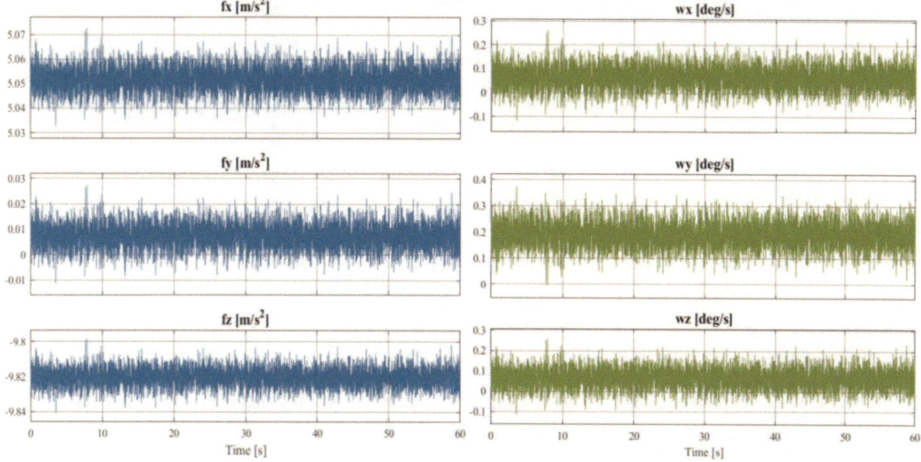

Figure 17.
Perturbed IMU outputs: Accelerations – Left, angular speeds – Right.

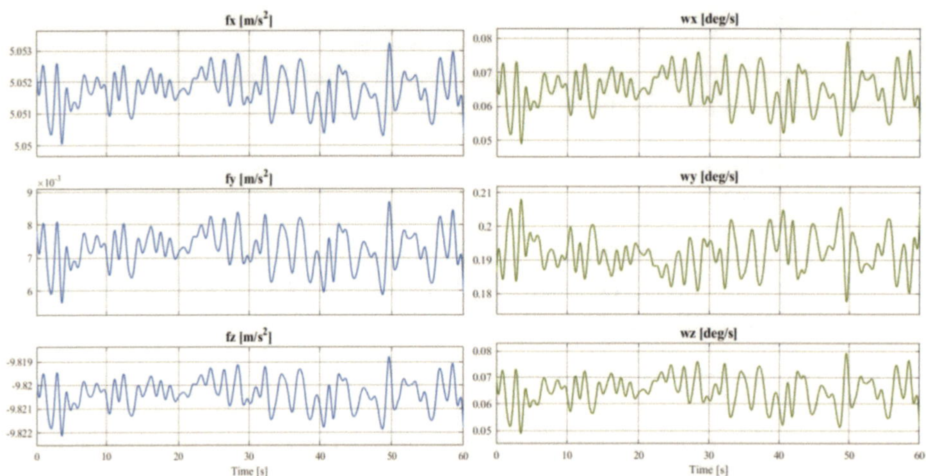

Figure 18.
Filtered IMU outputs: Accelerations – Left, angular speeds – Right.

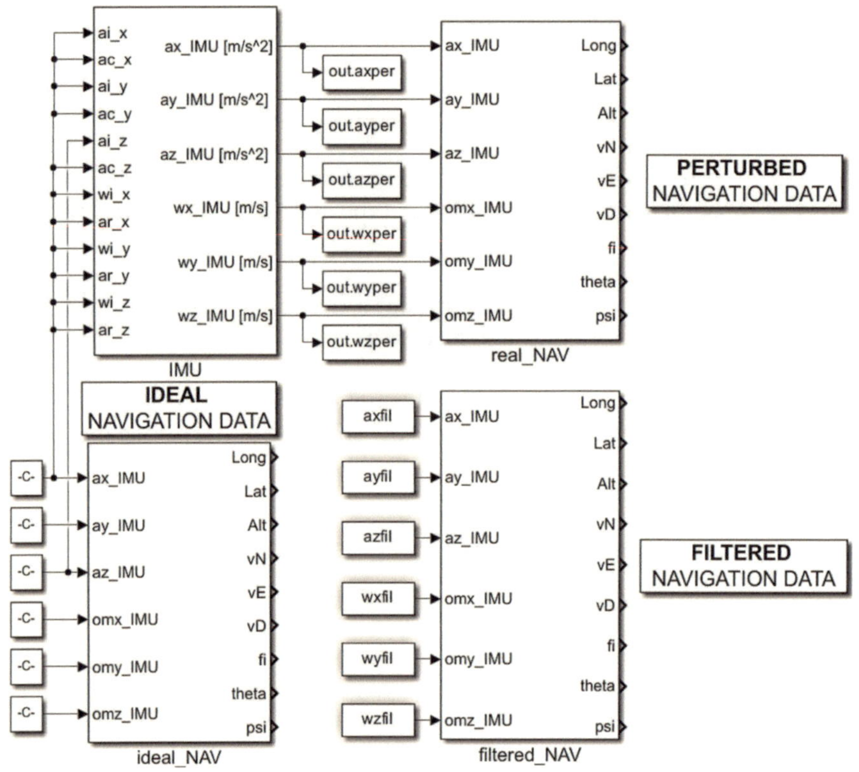

Figure 19.
MATLAB/Simulink configuration of the three algorithm scenarios.

- **Filtered Case:** The 'NAV' block utilizes the filtered inputs defined in **Figure 18**, obtained by denoising the outputs shown in **Figure 17**.

Figure 19 presents the MATLAB/Simulink implementation of the three simulation cases.

6.4 Results of the numerical simulation

The evaluation entailed comparing the absolute errors between the nine parameters (latitude, longitude, altitude, speed components in NED, and attitude angles) obtained from both the perturbed and filtered cases and the ideal navigation solution, which was illustrated in **Figure 20**. The evolution of the parameters follows the expected pattern from the picked numerical values:

- λ remains constant, Φ slightly increases, while h rises to almost 700 m from the initial value of 20 m;
- v_N increases, v_E remains null and v_D decreases;
- φ and ψ remain null, while θ increases to almost 12°.

Figure 21 shows the graphical representation of the errors obtained in both cases: red color for the perturbed case and blue color for the filtered case.

Right from the outset, it becomes evident that the error curve of the filtered case exhibits a less pronounced trend compared to the perturbed case. This visual observation suggests that the filtering procedure enhances the error evolution in relation to the ideal navigation solution. In terms of numerical analysis, **Table 4** presents the maximum absolute error values obtained for each parameter and input case.

In every parameter case, the errors consistently decrease after filtration. Notably, the most significant improvement was observed in the altitude parameter, which saw a reduction of nearly 60 meters following wavelet denoising. Additionally, the

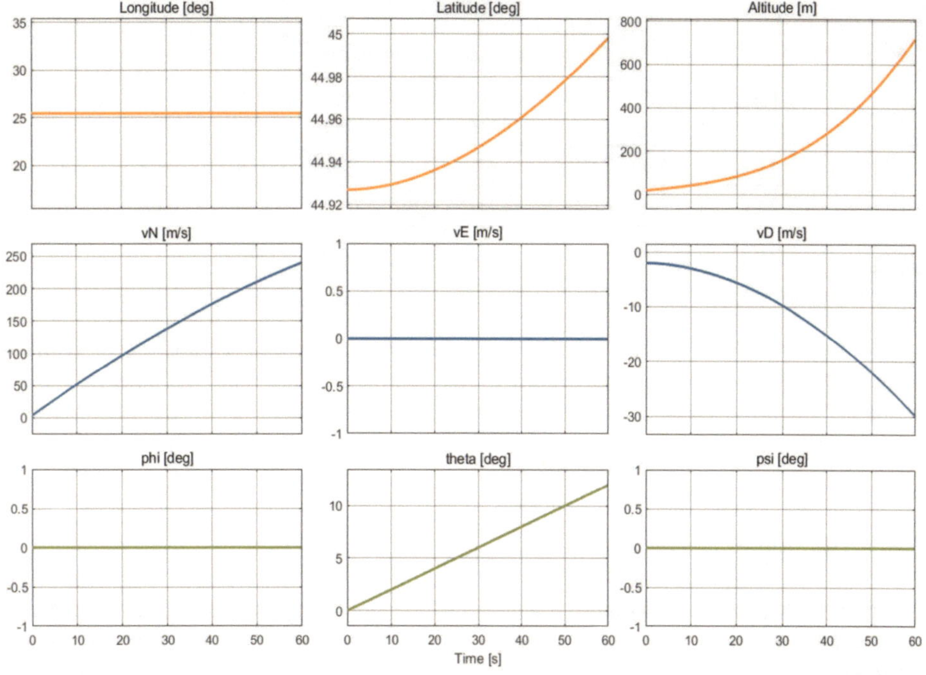

Figure 20.
Ideal navigation solution.

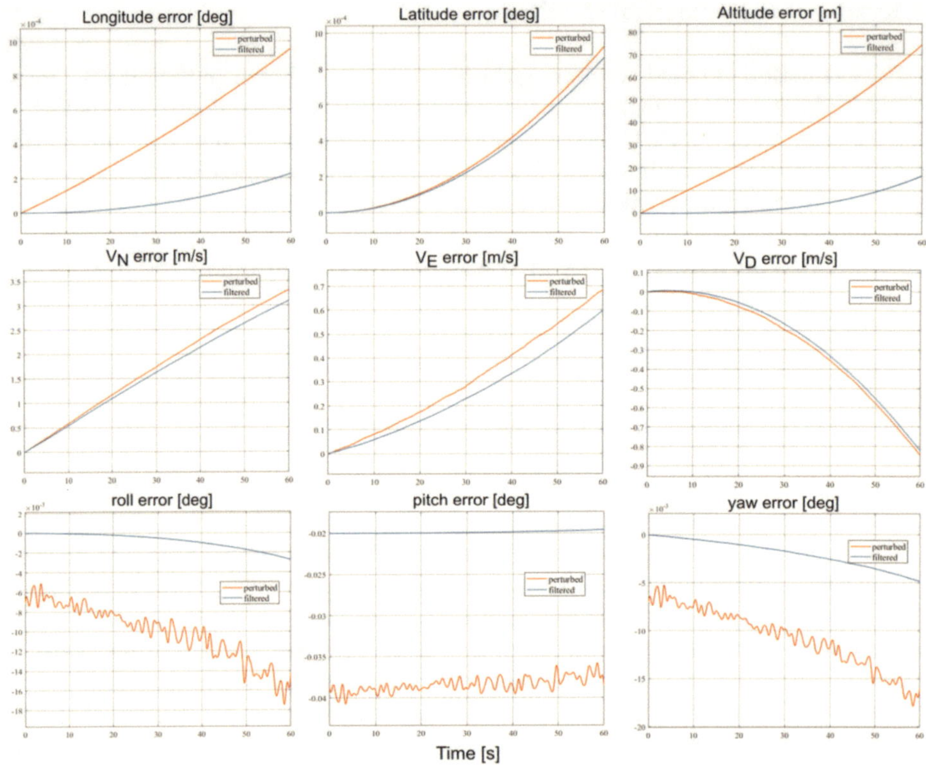

Figure 21.
Absolute errors obtained during the simulation.

Model	Maximum absolute error								
	λ [deg] $\cdot 10^{-4}$	Φ [deg] $\cdot 10^{-4}$	h [m]	v_N [m/s]	v_E [m/s]	v_D [m/s]	φ [deg] $\cdot 10^{-3}$	θ [deg]	ψ [deg] $\cdot 10^{-3}$
Perturbed	9.56	9.09	74.65	3.42	0.68	0.87	17.64	0.04	18.02
Filtered	2.18	8.47	16.78	3.18	0.58	0.83	2.76	0.02	5.01

Table 4.
Maximum absolute errors obtained during the simulation.

attitude angles exhibit a clear enhancement in signal characteristics, effectively alleviating the oscillatory noise introduced to the output signal of the MEMS sensors. The overall performance of the wavelet filter proves to be notably high, demonstrating a clear improvement in the positioning accuracy of the monitored vehicle.

7. Conclusions

The chapter presented a modern approach on improving the measurements of low-cost miniaturized inertial sensors, which currently stand as a viable technical avenue for the architecture of inertial measurement units. These sensors can be seamlessly incorporated into the framework of strap-down inertial navigation systems, catering to diverse applications like aerial vehicle surveillance. Consequently, the precision and positioning accuracy of these sensors emerge as crucial focal points for researchers to enhance.

The primary aim of this study was to evaluate the applicability of wavelets, a potent analytical tool, in addressing the challenge posed by perturbed navigation data in strap-down MEMS IMUs.

Our research was divided into two distinct phases. The initial phase encompassed a comprehensive exposition of miniaturized inertial sensors and their typical errors. Subsequently, the issue of signal corruption associated with these sensors was outlined, along with the prerequisites for its resolution. The theoretical foundation of wavelet theory was introduced, accompanied by an enumeration of key reasons that render it a suitable choice for the present case.

The first experimental part involved an examination of the wavelet filter's impact on an accelerometer and gyro model, both realized within MATLAB/Simulink. Distinct functions were employed for each sensor, and their efficacy was gauged through the computation of evaluation metrics. Following the application of the denoising technique using the MATLAB/Wavelet Analyzer tool, there was a notable reduction in noise levels, leading to enhanced signal attributes.

The second phase aimed at globally validating the efficacy of the wavelet filter by applying it to the complete strap-down navigation algorithm. The mathematical formulations of both attitude determination and position and velocity estimation components of the navigation algorithm were outlined, followed by their software implementation within MATLAB/Simulink. A practical scenario depicting the take-off of an aerial monitored vehicle was selected for numerical simulation of the strap-down algorithm. In this instance, to assess the performance of the denoising technique, three distinct input cases for the algorithm were examined. The ultimate objective was to compare the influence of unfiltered and filtered measurements on the navigation computation. Through a comparison of the absolute errors acquired in these two scenarios with the hypothetical ideal navigation solution, wherein numerical values serve as inputs, the outcomes underscored a substantial enhancement in positioning accuracy achieved through wavelet filtering. The evidence is presented through both graphical depictions and numerical values of the maximum absolute errors.

The overall conclusion drawn from the work presented in this chapter uncovers the potency of wavelets as a signal processing tool, offering remarkable attributes that hold promise for further advancements in the realm of MEMS-based strap-down INS. Given the numerous advantages bestowed by MEMS technology on the sensor market, the incorporation of wavelet filtering stands as a potentially advantageous step that could yield even greater benefits. The synergy between these two domains presents a compelling proposition that merits careful consideration and exploration.

Author details

Bianca-Gabriela Antofie[1], Marius-Ciprian Larco[1] and Teodor Lucian Grigorie[2*]

1 Military Technical Academy "Ferdinand I", Bucharest, Romania

2 National University of Science and Technology Politehnica Bucharest, Bucharest, Romania

*Address all correspondence to: ltgrigorie@yahoo.com

References

[1] Salychev O. Inertial Systems in Navigation and Geophysics. Moscow: Bauman MSTU Press; 1998

[2] Ramalingam R, Anitha G, Shanmugam J. Microelectromechnical systems inertial measurement unit error modelling and error analysis for low-cost strapdown inertial navigation system. Defence Science Journal. 2009;**59**: 650-658

[3] Edu IR, Adochiei FC, Obreja R, Rotaru C, Grigorie TL. New tuning method of the wavelet function for inertial sensor signals denoising. In: CSCC '14, Santorini Island, Greece, July 17-21, 2014. 2014

[4] Al Bassam N, Ramachandran V, Eratt PS. Wavelet theory and application in communication and signal processing. In: Wavelet Theory. London, UK: IntechOpen; 2021. DOI: 10.5772/intechopen.95047

[5] Pupo BL. Characterization of Errors and Noises in MEMS Inertial Sensors Using Allan Variance Method [Thesis]. Spain: Universitat Politècnica de Catalunya Barcelona Tech – UPC; 2016

[6] Grigorie TL, Lungu R. Traductoare Accelerometrice și Girometrice. Craiova: Sitech; 2005

[7] Nassar S, El-Sheimy N. Wavelet analysis for improving INS and INS/DGPS navigation accuracy. Journal of Navigation. 2005;**58**:119-134

[8] Wirsing K. Time frequency analysis of wavelet and Fourier transform. In: Wavelet Theory. London, UK: IntechOpen; 2021. DOI: 10.5772/intechopen.94521

[9] Hasan AM, Samsudin K, Ramli AR, Azmir RS. Wavelet-based pre-filtering for low cost inertial sensors. Journal of Applied Science. 2010;**10**:2217-2230

[10] Ismaeel S, Al-Khazraji A. Wavelet decomposition for Denoising GPS/INS outputs in vehicular navigation system. WSEAS Transactions on Systems. 2017; **16**:197-203

[11] Iontchev E, Miletiev R, Kapanakov P, Hristov L. Wavelet Algorithm for Denoising MEMS Sensor Data. Ohrid, North Macedonia: ICEST; 2019

[12] Arya SJ, Jisha VR, Usha PMK. Implementation and performance assessment of wavelet Prefiltered platform tilt computation using low-cost MEMS IMU. In: 2022 IEEE 1st International Conference on Data, Decision and Systems (ICDDS), Bangalore, India. Vol. 2. IEEE; 2022

[13] Song Q, Ma L, Cao J, et al. Image Denoising based on mean filter and wavelet transform. In: 2015 4th International Conference on Advanced Information Technology and Sensor Application (AITS). Epub ahead of print August 2015. 2015. DOI: 10.1109/aits.2015.17

[14] IEEE Standard for Inertial Sensor Terminology. IEEE Std 528–2001. IEEE Standard for Inertial Sensor Terminology; 2001. pp. 1-26

[15] Devices A. ADXRS150 ±150°/s Single Chip Yaw Rate Gyro with Signal Conditioning, Technical Datasheet. Analog Devices, Inc; 2004. pp. 1-12

[16] Silicon Designs, Inc. S1221x-025 Single-Axis Accelerometer, Technical Datasheet. Silicon Designs, Inc.; 2006. pp. 1-16

[17] Bayer FM, Kozakevicius AJ, Cintra RJ. An iterative wavelet threshold for signal denoising. Signal Processing. 2019;**162**:10-20

[18] Radix JC. Systèmes inertiels à composants liés “strap-down.”. Toulouse: Cepadues-Editions; 1993

[19] Grigorie TL, Sandu DG. On the attitude errors when the quaternionic Wilcox method is used. In: ELMAR 2007. Epub ahead of print September 2007. 2007. DOI: 10.1109/elmar.2007.4418804

[20] Farrell J, Barth M. The Global Positioning System and Inertial Navigation. New York: McGraw-Hill Education; 1999

[21] Dahia K. Nouvelles méthodes en filtrage particulaire: application au recalage de navigation inertielle par mesures altimétriques. 2005

[22] Grigorie TL. Echipamente de Bord și Navigație Aeriană–Îndrumar de laborator. Craiova: Sitech; 2013

[23] Grigorie TL, Sandu DG. Arhitecturi sinergice de sisteme de navigație cu componente inerțiale strap-down. Craiova: Sitech; 2013

Chapter 3

The Use of the 2D and 3D Complex Wavelet and Ridgelet Transforms in Geophysical Prospecting: Case of Potential Fields Data

Hassina Boukerbout

Abstract

The complex wavelet and ridgelet transforms are used in the potential field data interpretation for identifying the buried structures responsible for potential field anomalies. Its basis is the use of particular analyzing wavelets belonging to the Poisson semigroup that possess amazing properties regarding potential fields. In fact, the analyzed anomaly displays a conical signature in the wavelet domain and whose apex is pointing out at the causative homogeneous structure. Fundamentally, the interpretation is performed in the upward-continued domain where, the dilation of the wavelet transform is the upward-continuation altitude. This confers on the wavelet transform a considerable advantage: its robustness with respect to noise. The method is also developed to identify the depth, horizontal positions, size, strike direction, dips and shape of elongated 3D structures such as finite-dimensional dykes and faults. For this type of anomaly, the 2D wavelet transform corresponds to the ridgelet transform performed in the Radon domain, where elongated anomalies are recognized by high amplitude signatures. A reminder of the developed theory and applications in the 2D and 3D cases on real case studies are shown.

Keywords: continuous wavelet transform, ridgelet transform, radon domain, potential fields anomalies, 2D and 3D imaging

1. Introduction

One of the most important applications in geophysical prospecting is the identification, localization and characterization of bodies of geological interest, especially the causative sources of potential field anomalies (gravitational, magnetic, electrical, thermal and so on) measured at the surface or since high-resolution marine and airborne surveys are carried out. Thus, it continues to stimulate many important methodological developments in analysis and inversion techniques [1]. The purpose of inversion methods is to recover the source distribution using an integral equation that relates the measured potential field and the causative source distribution [2–5]. These analysis techniques contribute to reducing the non-uniqueness of the inverse problem by adding some a priori geological assumptions [6, 7]. Another category of processing methods that are not part of the inverse methods and which does not seek after the distribution of the source but can give information on the depth of the top of the causative sources [8, 9] or data transformation techniques such as downward and upward continuations, horizontal and vertical derivatives or reduction to the pole and oblique derivatives which produce the transformed fields, where the features of the original field are enhanced, using the relationship between the measured field and the distribution of its sources, based on a sum of convolution products using several transformation operators and an appropriate Green function [10–15]. However, these methods do not allow localizing sources along the z dimension, so another analyzing method based on the shape of the anomalies is the Euler deconvolution [16], which needs to be improved to eliminate noise effects since it is based on the calculation of gradients [17].

The continuous wavelet transform is an approach that can make easier the analysis of large amounts of data [18–21]. This method utilizes the homogeneity properties of the potential field, to identify and localize the causative sources [22–25]. Further works show the robustness of this method with respect to noise [26, 27], as revealed by many applications in geophysical prospecting, such as in aeromagnetics data [28–31], spontaneous electrical potential [32–34]; gravity data [35–37] and electromagnetic data [38]. The 2D wavelet method is then developed [30, 39] in order to localize and identify the potential fields anomalies causative structures in the case of elongated structures such as dykes, faults, etc. Thus, in the present work, after a brief recall of the theory in the cases 2D and 3D, an application to a part of aeromagnetic survey in the NW of Algeria will be presented and discussed.

2. The wavelet theory

The application of wavelets is recent in signal and image processing, but its mathematical history is much older since it basically works linking the Littlewood-Paley decomposition (1930), the version given by J.O. Strömberg from the basis of Franklin (1927) and the identity of Calderön (1960) [40].

Wavelet theory has experienced great development since the 1980s, to name a few references such as the work of Grossmann and Morlet (1984), [41, 42]. An interesting compilation of work on the developments and applications of wavelet theory for non-stationary signals in geophysics can be found in [19, 43, 44].

Wavelet analysis methods are essentially based on a representation of signals at different scales [45]. This is very interesting in geophysics since the information carried in the signals is carried by scaling laws or by non-stationaries. This is the case of seismological signals or potential fields whose variations represent the effects of multiscale or even fractal bodies of geological interest.

Other important developments have been made in wavelet theory since the work of Grossmann and Morlet [46], such as orthogonal wavelets, multiresolution analysis as well as the development of fast numerical algorithms [47–50]. The orthogonal wavelets are used to develop discrete wavelet transforms, unlike the continuous wavelet transform. An overview of how the discrete wavelet transforms can be used in the analysis of geophysical time series can be found in [51]. For instance, the continuous wavelet transform is used in the interpretation of potential field theory, and a review can be found in [28].

3. The continuous wavelet and ridgelet transforms

Here is a brief recall of the mathematical framework used in this work. The continuous wavelet transform [46] makes it possible to process large amounts of data. The wavelet transform allows for the detecting and characterizing homogeneous singularities in signals using the property of the homogeneity degree of the analyzed function [52–54].

3.1 The 1D continuous wavelet analysis of potential fields data

The wavelet transform w which is a convolution product of an analyzing wavelet g and a function φ_0, is defined as follows:

$$w[g,\varphi_0](b,a) \equiv \int_{\Re} \frac{1}{a} g\left(\frac{b-x}{a}\right)\varphi_0(x)dx = (D_a g * \varphi_0)(b) \tag{1}$$

$$D_a g(x) \equiv \frac{1}{a} g\left(\frac{x}{a}\right) \tag{2}$$

where D_a is the dilation operator, the dilation parameter $a > 0$, and b is the translation parameter.

To be admissible, the analyzing wavelet $g(x)$ should have a zero-mean oscillating behavior localized in a finite interval including the origin [19], which enables the wavelet transform to perform a local analysis of the signal [38]. In the case of complex continuous wavelet transform, this analyzing wavelet must be an oscillating complex function, localized on the real line [39]. The dilated wavelets $D_a g(x)$ are *a* then band-pass filters with bandwidth proportional to dilation a. This oscillating property means that the wavelet has a vanishing integral and allows the reconstruction of the analyzed signal from its wavelet transform [28, 38, 39]. Also, the covariance of the wavelet transform is an interesting property with respect to homogeneous functions [39], where the geometrical meaning is that the wavelet of homogeneous singularity displays a cone-like appearance whose apex points onto the singularity for $a \downarrow 0+$, so according to potential field theory, the analyzed field is produced by a homogeneous source located at (xs, zs) and, the dilation may correspond to an upward-continuation offset of the analyzed potential fields [24–27, 39]. So, the cone-like structure apex is then located at $(xs, a = -zs)$ below the positive half-plane of the wavelet.

3.2 The 2D continuous ridgelet transform of potential fields data

The 1D continuous wavelet transform method is applied to analyze 2D potential field anomalies data, measured in the horizontal plane by generalizing the Eq. (2) and then obtain the ridgelet transform [55, 56],

$$R[r,\varphi_0](b,a,s_\perp) = \int_{\Re^2} \frac{1}{a} r\left(\frac{b-x}{a}, y, s_\perp\right)\varphi_0(x,y)dxdy = \int_{\Re} \frac{dx}{a} g\left(\frac{b-x}{a}\right)\int_{\Re}\varphi_0(x,y)dxdy$$
$$= W\left[g, RT\left[\varphi_0, s_{//}\right]\right](b,a) \tag{3}$$

where $s_\perp$ a unit vector is perpendicular to the anomaly strike, $s_{//}$ is a unit vector in the direction of the elongated anomaly. The analyzing ridgelet is obtained by steering a 1D Poisson wavelet $g(x)$ in the perpendicular direction y, RT is the Radon transform of the potential field anomaly. The ridgelet transform of 2D anomalies, as defined in Eq. (3) is obtained by computing the wavelet transform for each value of the angular parameter in the Radon domain [28, 38, 39]. We use the complex ridgelet transform since we use complex analyzing wavelets [39]. The complex wavelet transform allows to analyze a signal using modulus and phase. The phase of the complex wavelet transform provides information about the inclination of the source. The imaginary part is obtained from the Hilbert transform of the real part. In this case, the complex wavelets correspond to analytical signal and their modulus and phase can be determined [27, 38].

Both the modulus and the phase of the ridgelet transform are used to localize the sources and, display conspicuous cone-like patterns associated with each analyzed anomaly. This cone is pointing to the source depth $\boldsymbol{zs}$ [39]. These locations are obtained by testing the geometry of the cone pointing towards the depth of the source for each grid point, in the half-space $(\boldsymbol{x}, \boldsymbol{z})$ [38, 39]. Subsequently, a statistical method is introduced, and the likelihood for the occurrence of an apex at source location $(\boldsymbol{x_s}, \boldsymbol{z_s})$ is evaluated by the maximum entropy criteria $\boldsymbol{\rho}$ [38, 39, 57]:

$$\rho(x_s, z_s) = \frac{\ln N + \sum_{i=1}^{N} h_i \ln h_i}{\ln N} \tag{4}$$

N corresponds to the number of grid points; hi correspond to the histogram values of the slopes along the modulus or phase lines forming the cone; $\boldsymbol{\rho}$ takes its values in the range [0, 1]. The result is a tomography map of the sources.

4. Application to aeromagnetic data

In this section, we show some results obtained by applying the complex wavelet and ridgelet analysis, to identify and localize structures responsible for aeromagnetic field anomalies, in the seismogenic Cheliff basin (NW of Algeria). The geological and seismotectonics framework can be found in [58, 59]. This region is one of the most seismically active zones of the western Mediterranean Sea [58, 60], related to the collision between African and European plates since the Upper Cretaceous [61]. The kinematic models derived from the Atlantic Ocean magnetic anomalies study have shown that this convergence is linked to a counter-clockwise rotation of Africa relative to Eurasia [61, 62]. The seismic activity in this area is directly associated with the plate boundary between Europe and Africa. This region is known for having been the site of two destructive earthquakes: 9 September 1954, earthquake of Orleansville with a magnitude of 6.5 and 10 October 1980, El Asnam one, with a 7.3 magnitude.

The aeromagnetic data used in this work resulted from the digitization of aeromagnetic maps issued from the aeromagnetic survey of Algeria. This survey was carried out by Aero Service Corporation between 1970 and 1974. The maps, digitized by the shape recognition method, were interpolated at the nodes of a 325 × 325 regular grid and reduced to a pole (**Figure 1**). This map displays important magnetic anomalies in North, along the coast, in the Mediterranean Sea and in the South. The ridgelet analysis results are shown in **Figure 2**. The 3D image obtained attests to the geological

DOI: http://dx.doi.org/10.5772/intechopen.1003873

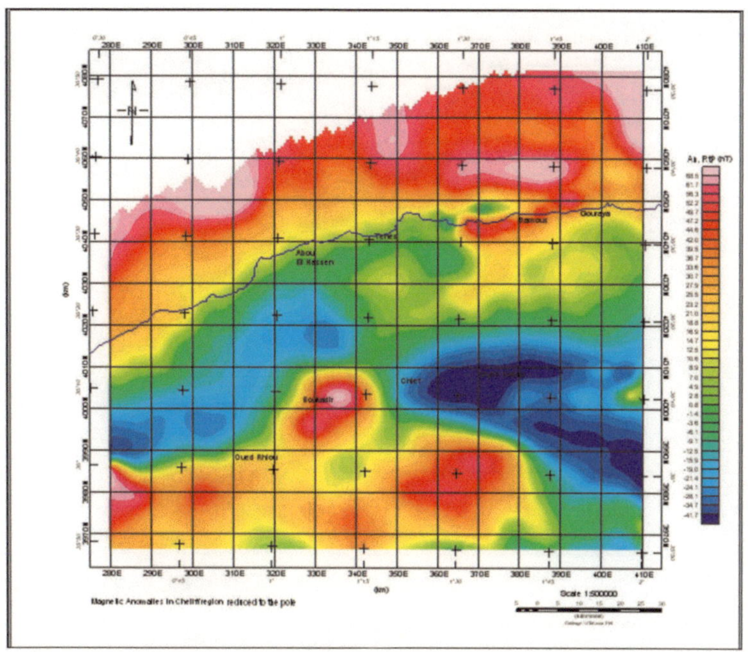

Figure 1.
Aeromagnetic map of Chlef region: total field anomaly reduced to the pole (Modified from [31]).

and seismotectonic complexity of the area. The elongated structures identified are a juxtaposition of prismatic bodies at different depths. The structures bordering the Chelliff basin are elongated in the E-W direction. In the North, in the Mediterranean Sea, their depth reaches 31 km. At the coast, within the volcanic structures, these structures reach a depth of 20 km. To the south of the basin, at the Ouarsenis Mountains, the depth of these structures reaches 29 km. In the Cheliff basin, the structures are oriented in NE-SW, NW-SE and E-W directions and located between near-surface and 25 km, while those oriented in N-S direction are at depth ranging between 9 and 16 km. Elongated structures oriented N-S appear to the North, limiting offshore and coastal anomalies. These structures reach depths of 20 km.

In order to skecth out the topography of the magnetized substratum and identify the structures in depth, the complex wavelet transform is applied to a N-S magnetic anomalies profile, located at 1°40′ and crossing magnetic anomalies from Mediterranean Sea to the North, as far as the Ouarsenis Mountains in South. **Figure 3** shows the intensity of magnetic anomaly (top of figure) varying from −40nT to +40nT. The middle of the figure corresponds to the modulus of the wavelet transform and the 2D image (bottom), where the magnetized substratum top depth, identified by the maximum entropy criteria, ranges between 6 and 30 km. The deepest is located in North (Med. Sea) and South (Ouarsenis Mounts); less deep in the sedimentary basin (thickness of basin). Many faults and contacts are identified along this profile. We can cite the Oued Fodda region (site of 1980 earthquake), where the O. Fodda fault is identified at the latitude of 4010 km at a depth of 6 km, with an inclination of 30° identified from the phase of the wavelet (**Figure 4**), these results are confirmed by the results of dislocation model of vertical movements for this area [59]. A good correlation is shown with N-S geological profile (**Figure 5**).

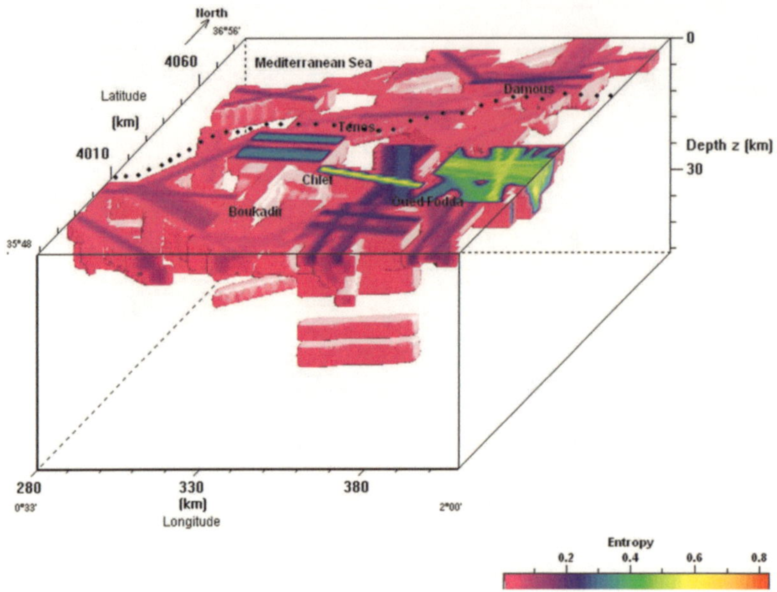

Figure 2.
3D imaging of magnetized structures identified from the complex ridgelet transform. The North direction is given by the latitude axis. The color scale corresponds to the maximum entropy criteria to select the source location (modified from [31]).

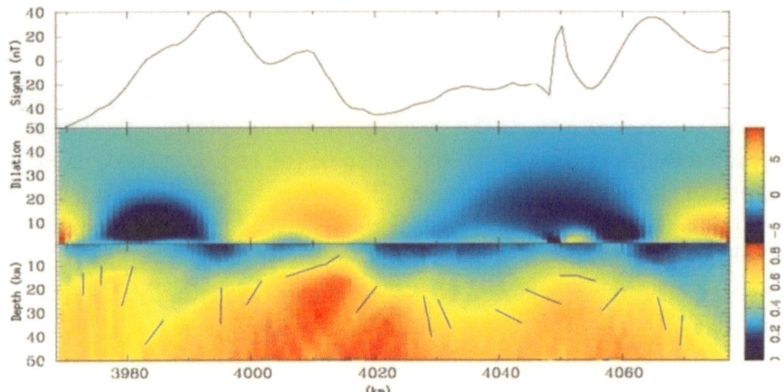

Figure 3.
The shape of the magnetized substratum and the identified structures in-depth, along a N-S profile (1°40′E) (modified from [31]).

5. Conclusion

The use of the wavelet and ridgelet transform in geophysical exploration allows for the identification of sources of potential field anomalies in 2D and 3D cases. The wavelet transform possesses many interesting mathematical properties with respect to potential fields theory; studies show that when applied to potential fields, it can have a deep physical sense, since the idea in the use of a homogeneous source is that an elongated geological structure may be replaced by a small number of equivalent point

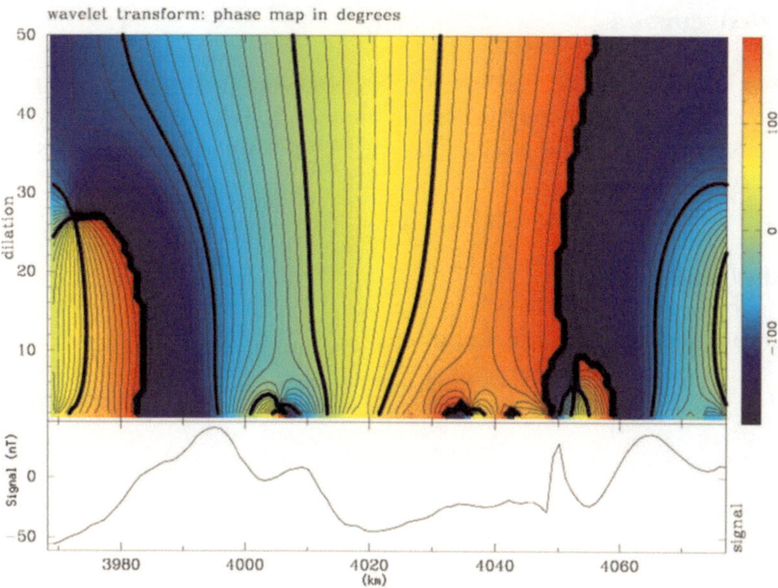

Figure 4.
Phase map of the complex continuous wavelet transform (modified from [31]).

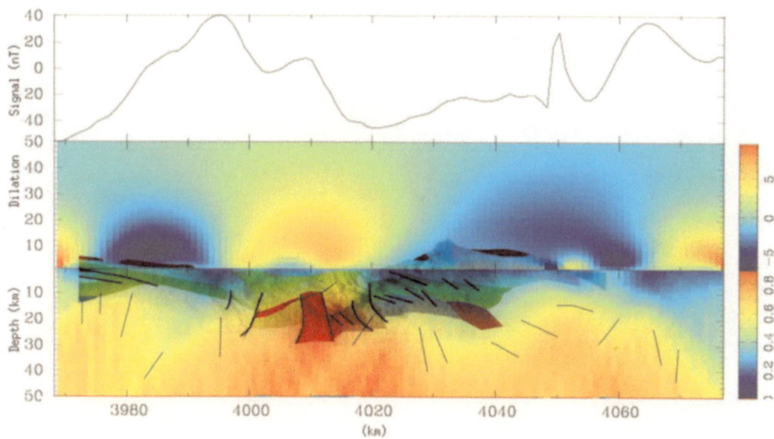

Figure 5.
Correlation with geological N-S profile (modified from [31]).

sources. The homogeneity of these point sources depends on the shape of the geological structure. Also, in the range of dilations, the signal-to-noise ratio is much better and makes the method more robust.

The ridgelet transform helps in the automatic detection of elongated structures in 3D, and the information provided by the wavelet transform concerning the identified sources, such as dip, depth, and dimensions, can be used to reduce the non-uniqueness of the inverse problem considerably. The application of real aeromagnetic data, without any a priori assumptions, shows a good correlation with known geological structures and identifies many more unknown structures.

Acknowledgements

This work is supported by the Centre de Recherche en Astronomie, Astrophysique & Géophysique. Our gratitude to D. Gibert, F. Moreau, P.G. Saracco, Sailhac and H. Beldjoudi. We are grateful to the editor and reviewers and special thanks to L. Divic and Tonči Lučić.

Conflict of interest

The author declares no conflict of interest.

Author details

Hassina Boukerbout
Centre de Recherche en Astronomie, Astrophysique and Géophysique, Algiers, Algeria

*Address all correspondence to: s.boukerbout@craag.dz

References

[1] Blakely RJ. Potential Theory in Gravity and Magnetic Applications. New York: Cambridge University Press; 1996. 441 p

[2] Cuer M, Bayer R. Fortran routines for linear inverse problems. Geophysics. 1980;**45**:1706-1719

[3] Tarantola A. Inverse Problem Theory. New York: Elsevier; 1987

[4] Parker RL. Geophysical Inverse Theory. Princeton, NJ: Princeton Univ. Press; 1994

[5] Li Y, Oldenburg DW. Fast inversion of large-scale magnetic data using wavelet transforms and a logarithmic barrier method. Geophysical Journal International. 2003;**152**:251-265

[6] Pilkington M. 3-D magnetic imaging using conjugate gradients. Geophysics. 1997;**62**:1132-1142

[7] Bosch M, Guillen A, Ledru P. Lithologic tomography: An application to geophysical data from the Cadomian belt of Northern Brittany, France. Tectonophysics. 2001;**331**:197-227

[8] Spector A, Grant FS. Statistical models for interpreting aeromagnetic data. Geophysics. 1970;**35**:293-302

[9] Green AG. Magnetic profile analysis. Geophysical Journal of the Royal Astronomical Society. 1972;**30**:393-403

[10] Paul MK, Data S, Banerjee B. Direct interpretation of two dimensional structural faults from gravity data. Geophysics. 1966;**31**:940-948

[11] Bhattacharyya BK. Design of spatial filters and their application to high-resolution aeromagnetic data. Geophysics. 1972;**37**:68-91

[12] Baranov W. Potential fields and their transformation in applied geophysics. Geoexplorer Monograph Series. 1975;**6**:121

[13] Gibert D, Galdéano A. A computer program to perform transformations of gravimetric and aeromagnetic surveys. Computational Geosciences. 1985;**11**:553-588

[14] Sowerbutts WTC. Magnetic mapping of the Butterton dyke: An example of detailed geophysical surveying. Journal of the Geological Society. 1987;**144**:29-33

[15] Pilkington M, Gregotski ME, Todoeschuck JP. Using fractal crustal magnetization models in magnetic interpretation. Geophysical Prospecting. 1994;**42**:677-692

[16] Thompson DT. EULDPH: A new technique for making computer-assisted depth estimates from magnetic data. Geophysics. 1982;**47**:31-37

[17] Mikhailov VA, Galdéano A, Diament M, Gvishiani A, Agayan A, Bogoutdinov S, et al. Application of artificial intelligence for Euler solutions clustering. Geophysics. 2003;**68**:168-180. DOI: 10.1190/1.1543204

[18] Arneodo A, Argoul F, Bacry E, Elezgaray J, Muzy J F. Ondelettes multifractales et turbulences, de l'AND aux croissances cristallines. Paris: Diderot Editeur; 1995

[19] Holschneider M. Waavelets: An Analysis Tool. Oxford, U.K: Clarendon; 1995

[20] Torresani B. Analyse continue par ondelettes. Paris: InterEditions & CNRS Editions; 1995

[21] Mallat S. A Wavelet Tour of Signal Processing. 2nd ed. Academic Press; 1999

[22] Alexandrescu M, Gbert D, Hulot G, Le Moûel JL, Saracco G. Detection of geomagnetic jerks using wavelet analysis. Journal of Geophysical Research. 1995;**100**:557-572

[23] Alexadrescu M, Gibert D, Hulot G, Le Moûel JL, Saracco G. Worldwide wavelet analysis of geomagnetic jerks. Journal of Geophysical Research. 1996; **101**:975-994

[24] Moreau F, Gibert D, Holschneider M, Saracco G. Wavelet analysis of potential fields. Inverse Problem. 1997;**13**:165-178

[25] Hornby P, Boschetti F, Horovitz FG. Analysis of potential field data in the wavelet domain. Geophysical Journal International. 1999;**137**:175-196

[26] Moreau F, Gibert D, Holschneider M, Saracco G. Identification of sources of potential fields with the continuous wavelet transform: Basic theory. Journal of Geophysical Research. 1999;**104**: 5003-5013

[27] Sailhac P, Gibert D. Identification of sources of potential fields with the continuous wavelet transform: Two-dimensional wavelets and multipolar approximation. Journal of Geophysical Research. 2003;**108**(B5):262. DOI: 10.1029/2002JB002021

[28] Sailhac P, Gibert D, Boukerbout H. The theory of the continuous wavelet transform in the interpretation of potential fields: A review. Geophysical Pro Specting. 2009;**57**(4):517-525

[29] Boschetti F, Therond V, Hornby P. Feature removal and isolation in potential field data. Geophysical Journal International. 2004;**159**:833-841. DOI: 10.1111/j1365-246X.2004.02293.x

[30] Boukerbout H. Analyse en ondelettes et prolongement des champs de potentiel. Développement d'une théorie 3-D et application en géophysique [thesis]. France: Rennes University I; 2004

[31] Boukerbout H, Abtout A, Gibert D, Henry B, Bouyahiaoui B, Derder MEM. Identification of deep magnetized structures in the tectonically active Chlef (Algeria) from aeromagnetic data using wavelet and ridgelet transforms. Journal of Applied Geophysics. 2018;**154**: 167-181. DOI: 10.1016/j.jappgeo.2018. 04.026

[32] Gibert D, Pessel M. Identification of sources of potential fields with the continuous wavelet transform: Application to self-potential profiles. Geophysical Research Letters. 2001;**28**: 1863-1866

[33] Sailhac P, Marquis G. Analytic potentials for the forward and inverse modeling of SP anomalies caused by subsurface fluid flow. Geophysical Research Letters. 2001;**28**:1851-1854

[34] Saracco G, Labazuy P, Moreau F. Localization of self-potential sources in volcano-electric effect with complex wavelet transform and electrical tomography methods for an active volcano. Geophysical Research Letters. 2004;**31**:L12610. DOI: 10.1029/ 2004GL019554

[35] Martelet G, Sailhac P, Moreau F, Diament M. Characterization of geological boundaries using 1-D wavelet transform on gravity data: Theory and application to theHimalayas. Geophysics. 2001;**66**:1116-1129

[36] Fedi M, Premiceri R, Quarta T, Villani AV. Joint application of continuous and discrete wavelet transform on gravity data to identify shallow and deep sources. Geophysical Journal International. 2004;**156**: 7-21. DOI: 10.1111/j1365-246X.2004. 02118.x

[37] Abtout A, Boukerbout H, Gibert D. Gravimetric evidences of active faults and underground structure of the Chlef seismogenic basin (Algeria). Journal of African Earth Sciences. 2014;**99**(2):363-373. DOI: 10.1016/j. jafrearsci.2014.02.011

[38] Boukerbout H, Gibert D, Sailhac P. Identification of sources of potential fields with the continuous wavelet transform: Application to VLF data. Geophysical Research Letters. 2003; **30**(8):1427. DOI: 10.1029/ 2003GL016884

[39] Boukerbout H, Gibert D. Identification of sources of potential fields with the continuous wavelet transform: Two-dimensional ridgelet analysis. Journal of Geophysical Research. 2006;**111**:B07104. DOI: 10.1029/2005JB004078

[40] Meyer Y, Roques S, Wavelet analysis and applications. Ed. Frontières, Gif-sur-Yvette. 1992

[41] Meyer Y. Ondelettes et opérateurs. Paris: Hermann; 1990

[42] Daubechies I. Orthonormal bases of compactly supported wavelets. Communications on Pure and Applied Mathematics. 1988;**XLI**:909-996

[43] Foufoula-Geogiou E, Kumar P. Wavelet analysis in geophysics: An introduction. Wavelet Analysis and Its Applications. 1994;**4**:1-43. DOI: 10.1016/ B978-0-08-052087-2.50007-4

[44] Mallat S. Une exploration des signaux en ondelettes. Editions de l'Ecole Polytechnique. 2000

[45] Meyer Y, Jaffard S, Rioul O. L'analyse par ondelettes. Pour la Science. 1987:28-34

[46] Grossmann A, Morlet J. Decomposition of Hardy functions into square integrable wavelets of constant shape. SIAM Journal of Mathematics. 1984;**15**:732-736

[47] Meyer Y, Orthonormal wavelets. . In: Combes JM, Grossman A, Tchamitchian P, editors. In Wavelets, Time-Frequency Methods and Phase Space. Berlin: Springer-Verlag; 1989. pp. 21-37

[48] Mallat S. Multiresolution approximation and wavelet orthonormal bases of $I^2(r)$. Transactions of the American Mathematical Society. 1989; **315**:69-88

[49] Duabechies I. Ten Lectures on wavelets. Philadelphia: SIAM; 1992

[50] Rioul O, Duhamel P. Fast algorithms for discrete and continuous wavelet transforms. IEEE Transformation on Information Theory. 1992;**38**:569-586

[51] Percival DB. Analysis of geophysical time series using discrete wavelet transforms: An overview. In: Donner RV, Barbosa SM, editors. Nonlinear Time Series Analysis in the Geosciences. Lecture Notes in Earth Sciences. Vol. 112. Berlin, Heidelberg: Springer; 2008

[52] Grossmann A, Holschneider M, Kronland-Martinet R, Morlet J. Detection of abrupt changes in sound signals with the help of wavelet transforms. In: Advances in Electronics

and Electron Physics. Vol. 19. New York: Elsevier; 1987. pp. 289-306

[53] Holschneider M. On the wavelet transformation of fractal objects. Journal of Statical Physics. 1988;**50**:953-993

[54] Mallat S, Hwang WL. Singularity detection and processing with wavelets. IEEE Transactions on Information Theory. 1992;**38**:617-643

[55] Candès M. Theory and applications [thesis]. Stanford, CA: Dep. Of Stat. Stanford Univ.Dep. Of Stat. Stanford Univ.; 1998

[56] Candès M, Donoho DL. Ridgelets : A key to higher-dimensional intermittency? Philosophical Transactions on Royal Society Series A. 1999;**357**:2495-2509

[57] Tass P, Rosemblum MG, Weule J, Kurths J, Pikovsky A, Volkmann J, et al. Detection of n:M phase locking from noisy data: Application to magneto encephalography. Physical Review Letters. 1999;**81**:3291-3294

[58] Meghraoui M, Cisternas A, Philip H. Seismotectonics of the lower Chelif basin: Structural background of the El Asnam (Algeria) earthquake. Tectonics. 1986;**5**(6):809-836

[59] Bezzeghoud M, Dimitrov D, Ruegg JC, Lammali K. Faulting mechanism of the El Asnam (Algeria) 1954 and 1980 earthquakes from modelling of vertical movements. Tectonophysics. 1995;**249**:249-266

[60] Le Pichon X, Bergerat F, Roulet MJ. Plate kinematics and tectonic leading to alpine belt formation: A new analysis, processes in continental lithospheric deformation. Geological Society of America Special Paper. 1988;**218**:111-131. DOI: 10.1130/SPE218-p111

[61] Dewey JW. The 1954 and the 1980 Algerian earthquakes : Implications for the characteristic-displacement model of fault behavior. Bulletin of the Seismological Society of America. 1991; **81**(2):446-467

[62] Rosenbaum G, Lister GS, Duboz C. Relative motions of Africa, Iberia and Europe during Alpine orogeny. Tectonophysics. 2002;**359**(1-2):117-129

Chapter 4

Perspective Chapter: Detecting Volatility Pattern of Assets Returns Using Wavelet Analysis

Okonkwo Chidi Ukwuoma, Ugo Donald Chukwuma and Titus Ifeanyi Chinebu

Abstract

This chapter advocates for the use of wavelet analysis as a potent tool in understanding the dynamic nature of asset price volatility in financial markets. While traditional methods like GARCH models have been valuable, wavelet analysis offers a distinctive approach by decomposing time series data into various scales and frequencies. This enables a comprehensive perspective, capturing both short-term fluctuations and long-term trends. In an era of interconnected and information-rich financial markets, the ability to discern subtle volatility patterns is crucial. The chapter provides a guide to wavelet analysis, explaining its foundations, principles, and methodology for application to financial time series. Real data from NASDAQ Composite, DOW Incorporated, S&P500, and Omnicell Inc. is used for illustration. The efficacy of wavelet analysis is emphasized, offering finance professionals, academia, and researchers a simple yet robust approach to navigate the complexities of modern financial markets, make informed decisions, and adapt to evolving conditions. The chapter aims to enhance understanding of financial market behavior, inspiring further research and innovation in financial analysis and risk management.

Keywords: volatility, wavelet analysis, wavelet coherence, continuous wavelet transform, asset returns, stock market

1. Introduction

The dynamic landscape of the financial markets is very volatile, and the concept of volatility plays a pivotal role in the financial market [1, 2]. Volatility, often referred to as the degree of variation in asset prices over time, serves as a fundamental indicator for investors, traders, and risk managers [3, 4]. Accurate assessment of volatility patterns is important in making informed decisions, managing risk exposure, and optimizing investment strategies [5]. While traditional methods of volatility measurement, such as the generalized autoregressive heteroskedasticity (GARCH) models, have provided valuable insights, the complexity and nuances of market behaviors demand more alternative and more comprehensive approaches to volatility analysis [6–9]. In [10], the authors analyzed the multifractal properties of the US and European stock markets to identify patterns in investor sentiment.

This chapter seeks to look at an alternative approach to volatility analysis through the use of wavelet analysis. Wavelet analysis is a powerful mathematical tool that allows us to explore the temporal dynamics of volatility patterns with a level of detail that traditional methods cannot easily achieve. By decomposing financial time series data into different scales and frequencies, wavelet analysis offers a richer perspective on market behavior that captures both short-term fluctuations and long-term trends [11–13].

As the financial landscape becomes increasingly interconnected and information-rich, the ability to detect and interpret subtle volatility patterns becomes important. The chapter aims to provide a comprehensive understanding of wavelet analysis as a tool for uncovering these patterns and extracting meaningful insights from complex financial data. We will explore the foundations of volatility measurement, discuss the principles of wavelet analysis, delve into the methodology of applying wavelets to financial time series, and exemplify the approach using real data from the stock market.

Through this exploration, readers will gain a deeper appreciation of the power of wavelet analysis in understanding the intricate dynamics of asset returns volatility. By embracing this simple yet robust approach, finance, academia, and research professionals will be better equipped to navigate the complexities of modern financial markets, make informed decisions, and adapt to evolving market conditions.

2. Volatility in the financial market

The pattern of volatility in asset returns can provide insights into the risk and expected returns of the asset. The knowledge of volatility pattern is a great tool for investors as it helps them to know when to buy, sell, or hold their securities [1, 2, 6, 14]. There are certain stylized facts about assets return volatility. They are:

i. Higher return volatility increases the probability of a bear market.

ii. Stocks with high volatility risk tend to have higher expected returns.

iii. When volatility increases, the equity and variance risk premiums fall or stay flat at short horizons despite the higher future risk.

iv. Volatility clustering is a well-known stylized feature of financial asset returns, and there may be an asymmetric pattern in volatility clustering.

v. Increases in volatility positively forecast the variance risk premium at longer horizons.

3. Time domain and frequency domain analysis

Time domain analysis focuses on analyzing signals or mathematical functions in reference to time. It displays the changes in a signal over a span of time. It is commonly visualized using graphs or plots that show the signal's behavior over time. It provides insights into the temporal characteristics and dynamics of the signal. It is useful for studying the behavior of asset returns over different time periods. It can capture trends, patterns, and fluctuations in the returns over time [4, 5].

Frequency domain analysis, on the other hand, analyzes signals or mathematical functions in reference to frequency. It provides information about the distribution of signal energy across different frequencies. It is commonly visualized using tools, such as spectrum analyzers or frequency response plots. It helps to identify the presence of specific frequencies or frequency components in the signal. It is useful for studying the periodicity, cycles, and spectral characteristics of asset returns. It can reveal information about the dominant frequencies or frequency bands that contribute to the returns [14, 15].

3.1 Key differences between time and frequency domain

Representation: Time domain analysis represents signals in the time dimension, while frequency domain analysis represents signals in the frequency dimension.

Visualization: Time domain analysis is commonly visualized using graphs or plots of signal amplitude versus time, while frequency domain analysis is visualized using tools, such as spectrum analyzers or frequency response plots.

Focus: Time domain analysis focuses on the temporal behavior and changes in the signal, while frequency domain analysis focuses on the distribution of signal energy across different frequencies.

Insights: Time domain analysis provides insights into the temporal characteristics and dynamics of the signal, while frequency domain analysis provides insights into the spectral characteristics and frequency components of the signal.

Applications: Time domain analysis is useful for studying trends, patterns, and fluctuations over time, while frequency domain analysis is useful for studying periodicity, cycles, and spectral properties.

In the context of asset returns, time domain analysis helps understand the temporal behavior and patterns of returns, while frequency domain analysis helps identify dominant frequencies or cycles that contribute to the returns. Both approaches have their own advantages and can provide valuable insights into the characteristics of asset returns.

4. Wavelet analysis

Wavelet analysis is a tool that utilizes wavelets, which are mathematical functions to represent a signal in a localized and adaptable manner. It has the ability to capture the signal dynamics and identify the patterns and features in both time and frequency domains.

The capacity to evaluate time series data with nonstationary (changing over time) characteristics is a key benefit of wavelet analysis. This makes it practical for a variety of applications, including denoising, signal compression, and image and audio compression.

Numerous applications, including banking, biomedicine, engineering, and geophysics, have used wavelet analysis. It has been used in finance to evaluate financial time series data, including stock prices, to spot market patterns and trends. It can be used in biomedicine to examine physiological signals, such as brain activity and electrocardiograms. It can be used to analyze data in geophysics, such as seismic waves.

The wavelet transform takes a signal and changes it into a form that brings up certain features of the series for analysis. It is a real-valued and square-integrable function $\psi \in L^2(\mathbb{R})$ that satisfies the condition

$$\psi_{\tau,s}(t) = \frac{1}{\sqrt{s}}\psi\frac{(t-\tau)}{s}, s \in \mathbb{R}^+, \tau \in \mathbb{R}. \tag{1}$$

Where s in Eq. (1) is the scaling or dilation parameter and τ is a translation or position parameter. For a time series $x(t)$, its wavelet is oscillatory (goes up and down) that is $\int_{-\infty}^{\infty} \psi(x(t))dx = 0$. It is integrable, that is, $\int_{-\infty}^{\infty} |\psi(x(t))|dx < \infty$. It also satisfies the admissibility condition.

A wavelet $\psi \in L^2(\mathbb{R})$ is said to be admissible if its Fourier transform, $\mathrm{F}(x(t)) = \int_{-\infty}^{\infty} \psi(u)e^{-i2\pi ux(t)}du$, satisfy $C_\psi = \int_{-\infty}^{\infty} \frac{|F(x(t))|^2}{x(t)}dx$ where $0 < C_\psi < \infty$.

4.1 Continuous wavelet transform (CWT)

The CWT is given as

$$W_x(u,s) = \int_{-\infty}^{\infty} x(t)\frac{1}{\sqrt{s}}\psi\left(\frac{t-u}{s}\right)dt \tag{2}$$

In Eq. (2), $W_x(u,s)$ is simply a projection of a chosen wavelet on a time series. It is also the wavelet coefficient of the time series x. It is possible to reconstruct the decomposed CWT to recover the time series using

$$x(t) = \frac{1}{C_\psi}\int_0^{\infty}\left[\int_{-\infty}^{\infty} W_x(u,s)\psi_{u,s}(t)du\right]\frac{ds}{s^2}, s \neq 0 \tag{3}$$

Provided C_ψ satisfies the admissibility condition. Continuous wavelet transforms have a very high computational cost.

4.2 Wavelet coherence

Coherence is analog to classical correlation. To identify both frequency bands and time intervals when two signals are related, wavelet coherence is used. Given signal X (t); Y(t), their wavelet coherence is defined as

$$R_{x,y}^2(s,\tau) = \frac{|S(s^{-1}W_{x,y}(s,\tau))|}{\sqrt{S\left(s^{-1}|W_x|^2\right) \times S\left(s^{-1}|W_y|^2\right)}}, 0 \le R_{x,y}(s,\tau) \le 1 \tag{4}$$

where S in (4) is a smoothing operator defined as $\mathrm{S(W)} = S_{\mathrm{scale}}(S_{\mathrm{time}}(\mathrm{W(s)}))$, where S_{scale} denotes smoothing along the wavelet scale axis and S_{time} smoothing in time [16]. and is dependent on the choice of mother wavelet. Values of the wavelet coherence close to zero indicate weak correlation, while values close to one show evidence of strong correlation.

5. Implementation of wavelet analysis in R

Several R packages are available for the implementation of wavelet analysis in R. Some of the packages and their descriptions:

1. WaveletGARCH: This package is used to fit the Wavelet-GARCH Model to Volatile Time Series Data.

2. Wavelets—This package provides functions for computing wavelet filters, wavelet transforms, and multiresolution analyses. It also includes functions for plotting wavelet transform filters.

3. WaveletComp—This package provides functions for wavelet analysis and reconstruction of time series, cross-wavelets, and phase difference. It also includes functions for significance with bootstrap algorithms.

4. Waveslim—This package provides functions for wavelet-based signal processing, including denoising, compression, and signal reconstruction. It also includes functions for wavelet-based time series analysis and visualization.

5. Wavethresh—This package provides functions for wavelet thresholding, including hard, soft, and adaptive thresholding. It also includes functions for wavelet-based denoising and signal reconstruction.

6. WaveletML: This package decomposes time series into different components which helps to capture volatility at multi resolution level by various models. Then it uses Machine Learning models (Artificial Neural Network and Support Vector Regression have been used) is used for data predictions.

7. WaveletANN: This package wavelet and ANN technique to de-noise data and make forecast.

8. Biwavelets—This package provides functions for wavelet analysis for univariate and bivariate wavelet analysis.

These R packages provide a range of functions for wavelet analysis, including continuous wavelet transform, wavelet coherence, wavelet cross-spectrum, wavelet packet transforms, and wavelet variance stabilization. They also provide functions for plotting wavelet transform filters and significance with bootstrap algorithms. These packages can be useful for detecting volatility patterns in financial returns and analyzing signals or functions in both the time and frequency domains.

6. Algorithm for implementation of wavelet analysis

i. Collect data

ii. Clean the data

iii. Compute the descriptive statistics

iv. Get the returns of the data

v. Ensure the length of the data is of the power 2^n

vi. Compute and plot the continuous wavelet transform of the returns of the stock

vii. vii Compute and plot the wavelet coherence

viii. Interpret the result

7. DATA and data analysis

The data utilized for the example are from NASDAQ Composite (NASDAQ), DOW Incorporated (DOW), S&P500 (SP500), and Omnicell Inc. (OMCL). Sourced from Yahoofinance.com from August 8, 2019 to September 1, 2023. We first compute the stock price returns and then follow all the steps in the algorithm as outlined above.

8. Result and interpretation of result

From the summary statistics in **Table 1**, it can be seen that the standard deviation measures the dispersion of the data around the mean. The higher the standard deviation, the more volatile the instrument is. Looking at the standard deviation of the four stocks, we can see that DOW has the highest standard deviation of 5881.868, followed by NASDAQ with 3312.895, SP500 with 828.6003, and OMCL with 38.96592. Therefore, we can say that DOW is the most volatile stock, followed by NASDAQ, SP500, and OMCL. In **Table 2**, we observe that the standard deviation of the returns shows that the OMCL returns are the most volatile followed by NASDAQ, SP500, and DOW. These results show that a stock could be highly volatile while its returns are not.

The time series plot in **Figures 1–4** indicates that market forces were dominant, as evidenced by the movement of the stocks. They all reached their lowest point simultaneously, as shown by the light blue oval highlighter. In the same vein, they also

	SP500	DOW	NASDAQ	OMCL
Mean	3160.181	26664.38	9182.505	74.36086
Median	2919.35	26213.1	8070.12	67.21
Mode	4354.17	17924.24	7208.17	43.05
Std Deviation	828.6003	5881.868	3312.895	38.96592
Kurtosis	−1.22743	−1.15351	−1.17895	0.121511
Skewness	0.29609	−0.07548	0.354491	0.990368
Minimum	1833.4	15691.62	4218.81	25.45
Maximum	4804.51	36722.6	16120.92	185
Count	2011	2011	2011	2011

Table 1.
Summary statistics of the stock price.

	SP500r	DOWr	NASDAQr	OMCLr
Mean	0.000437	0.000284	0.000565	−0.0002
Median	0.001191	0.000756	0.001913	0.000177
Mode	0.011773	0.00956	0.009498	0
Std_Dev	0.011856	0.011647	0.016103	0.027882
Kurtosis	6.800708	14.20334	5.781579	27.42
Skewness	−0.50498	−0.58657	−0.55666	−2.01229
Minimum	−0.07141	−0.10353	−0.09504	−0.34406
Maximum	0.060343	0.081574	0.068863	0.140509

Table 2.
Summary statistics of the stock price return.

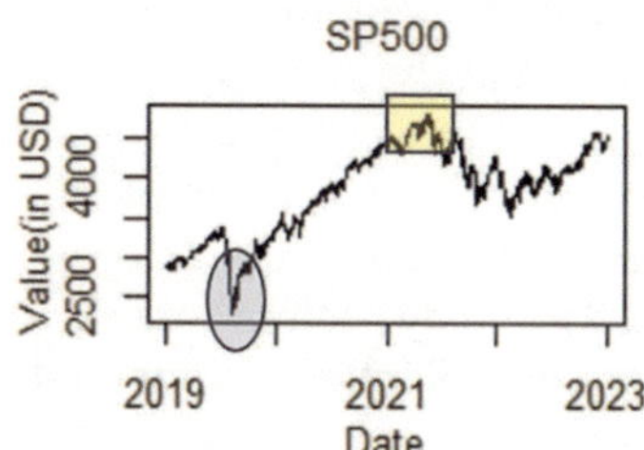

Figure 1.
Time series plot of SP500.

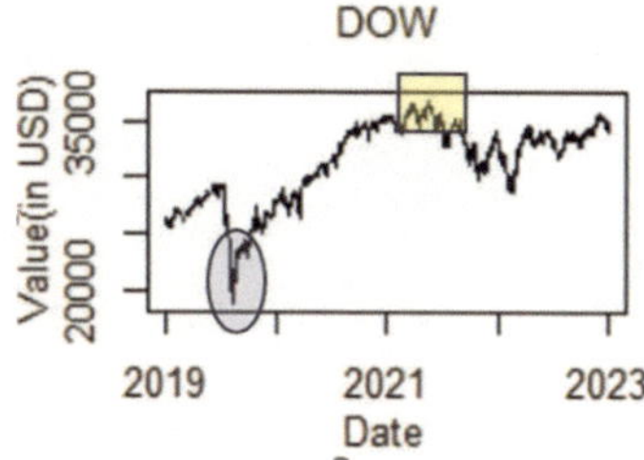

Figure 2.
Time series plot of DOW.

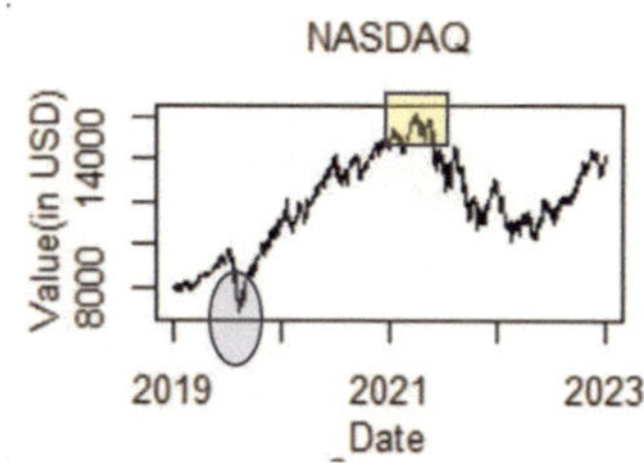

Figure 3.
Time series plot of NASDAQ.

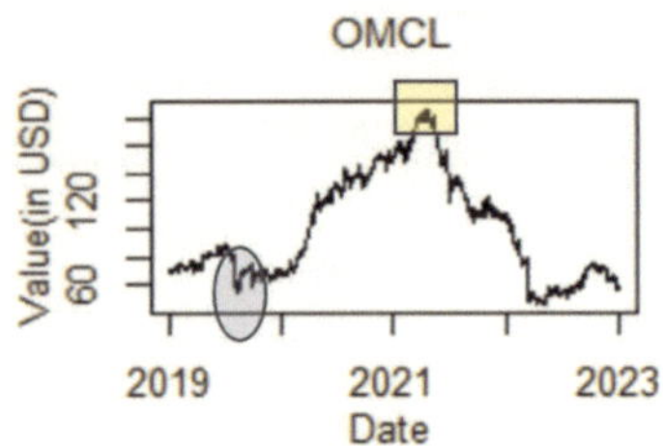

Figure 4.
Time series plot of OMCL.

experience their peak at the same period as highlighted by the light-yellow rectangles. In between these two periods, the stock price rose consistently until it reaches its peak, they all took a downward move. While the others were struggling to stay afloat, OMCL took a nose dive to a very low point (**Figures 5–8**).

The stock returns indicate that the point of highest volatility (with the highest width shown by the green oval) occurred in the same period for all stock returns except OMCLr whose highest volatility occurs at the point shown by the pink highlight.

The continuous wavelet transform (cwt) attempts to capture the volatility of the stock return in time and frequency domain. The color indicates the intensity of the volatility in any given region. The colder colors (blue through green) indicate regions of low volatility, while the hot colors (yellow through red) indicate regions of high volatility. The x-axis captures the time domain while the y-axis(period) captures the frequency domain. The frequency domain goes from top (0) to bottom (256). The area covered by the cone-like shape is called the cone of influence. This is the region in which the values are statistically significant.

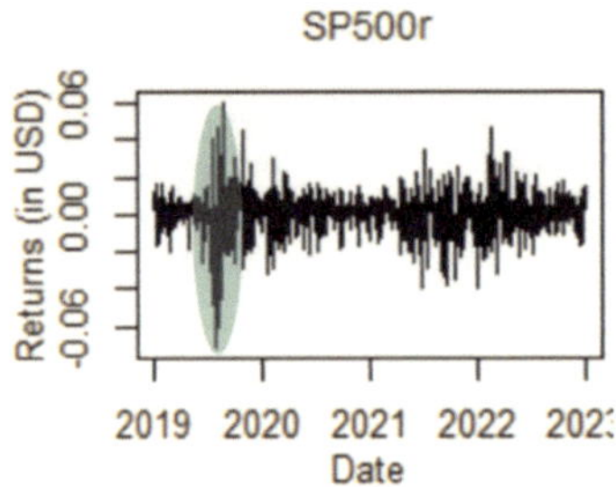

Figure 5.
Time series plot of SP500. Returns.

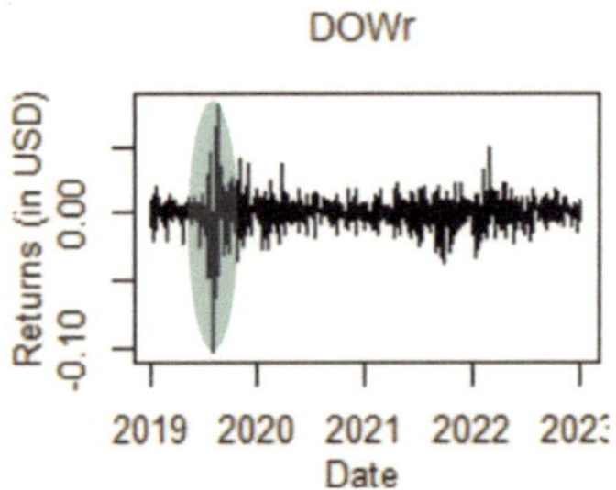

Figure 6.
Time series plot of DOW returns.

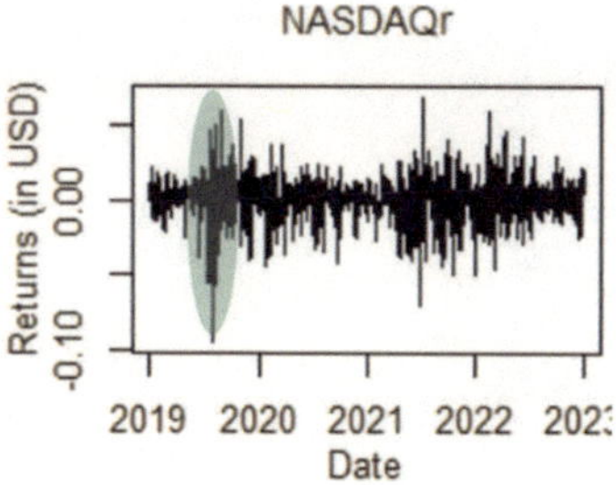

Figure 7.
Time series plot of NASDAQ returns.

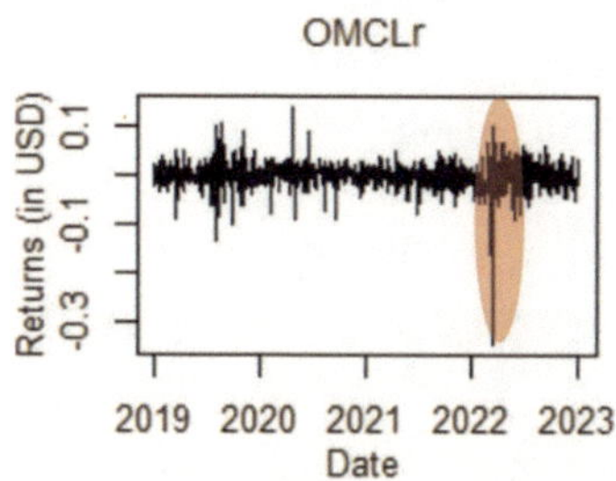

Figure 8.
Time series plot of OMCL returns.

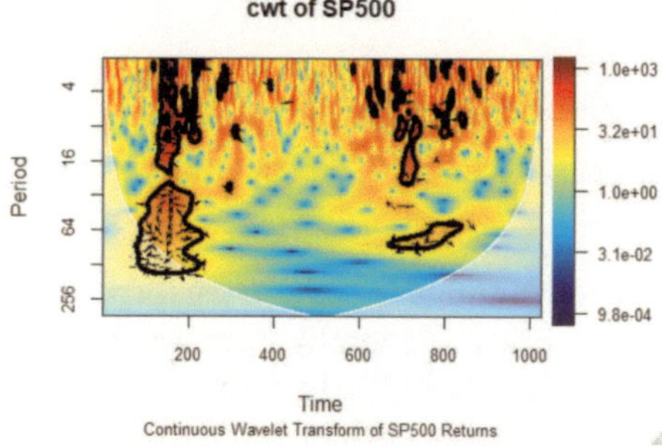

Figure 9.
Continuous wavelet transform of SP500 returns.

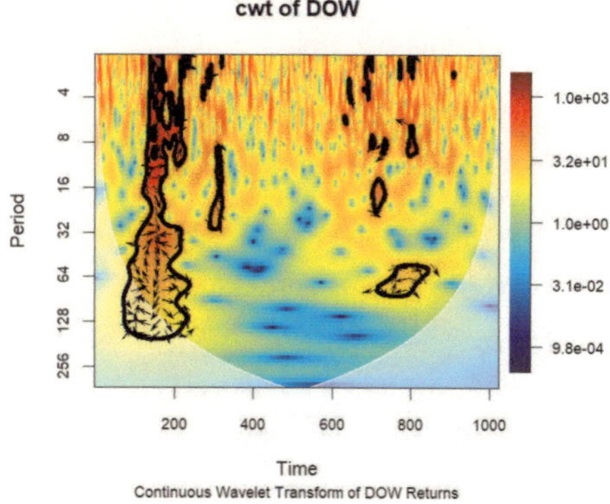

Figure 10.
Continuous wavelet transform of DOW returns.

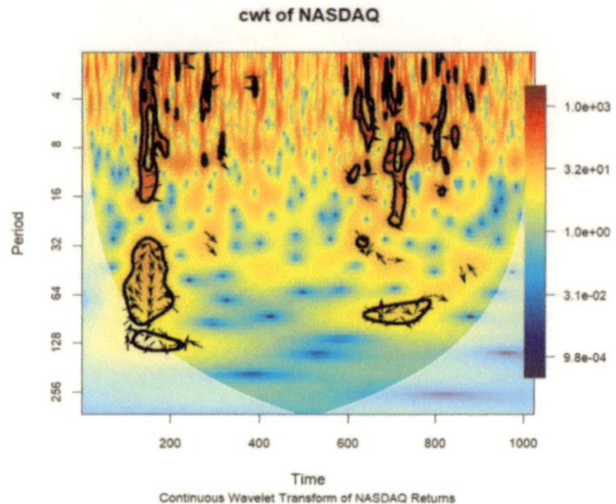

Figure 11.
Continuous wavelet transform of NASDAQ returns.

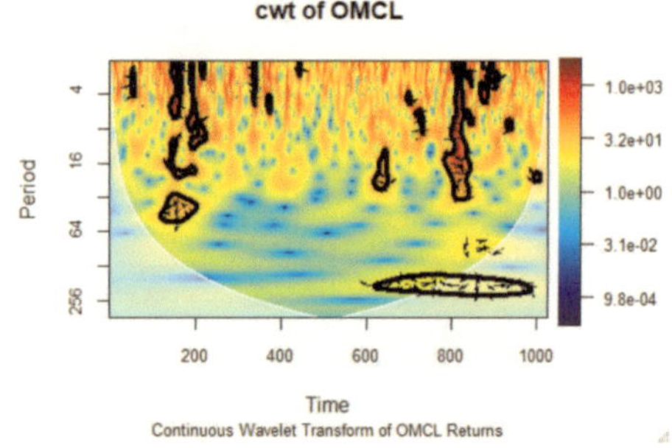

Figure 12.
Continuous wavelet transform of OMCL returns.

The cwt plots **Figure 9** Continuous wavelet transform of SP500 returns. **Figures 9–12** shows that the returns generally have a very high volatility in the low-frequency domain and low volatility in the high-frequency domain. The NASDAQ return seems to have a higher volatility spread at higher frequencies. This is followed by OMCL.

The wavelet coherence (wtc) captures the comovement of two securities in time and frequency domain. It shows well their returns move together. The cooler color

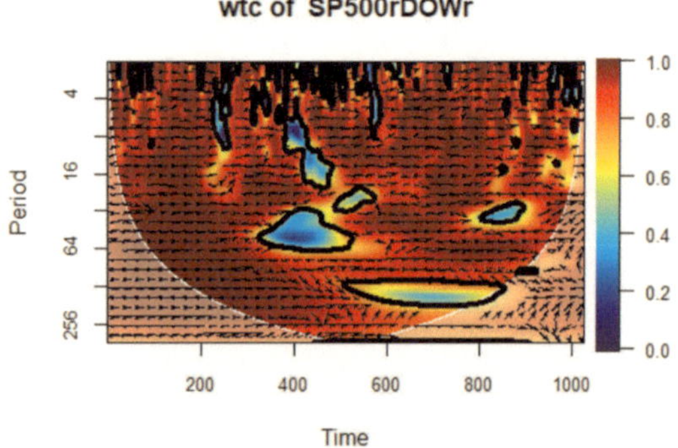

Figure 13.
Wavelet coherence of SP500r and DOWr.

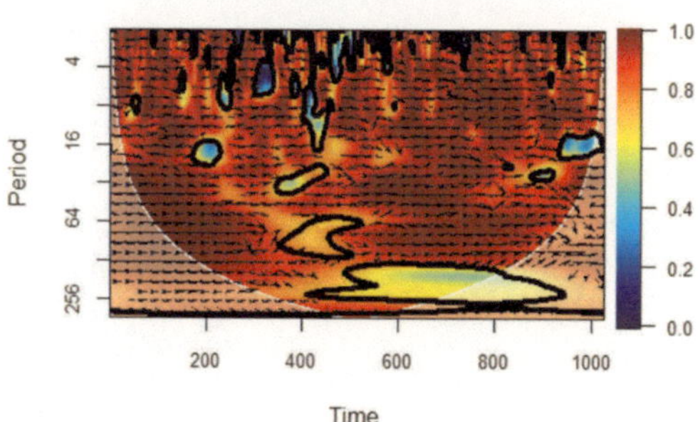

Figure 14.
Wavelet coherence of SP500r and NASDAQr.

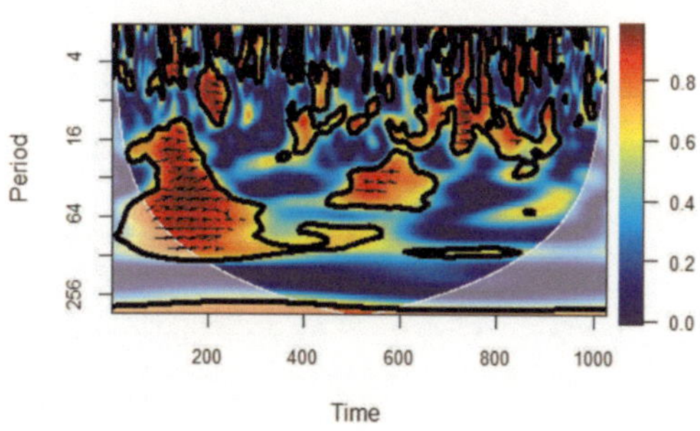

Figure 15.
Wavelet coherence of SP500r and OMCLr.

(blue) shows that the two returns have low comovement, while the hotter color(red) shows a very high comovement.

The wavelet coherence plots show that there is a very strong comovement between SP500r and DOWr **Figure 13**, as well as between SP500r and NASDAQr **Figure 14** at almost all time and frequency (**Figure 15**). The comovement of DOWr and NASDAQr **Figure 16** is very high except some island between the 250 and 600 in the time domain where there are interspersed with low volatility. The comovements of SP500r and OMCLr, DOWr and OMCLr, as well as NASDAQr and OMCLr are generally low with flashes of spots with high volatility. The implication of this for

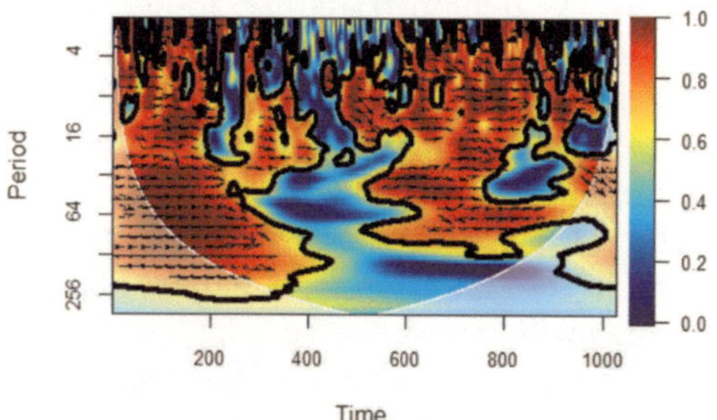

Figure 16.
Wavelet coherence of DOWr and NASDAQr.

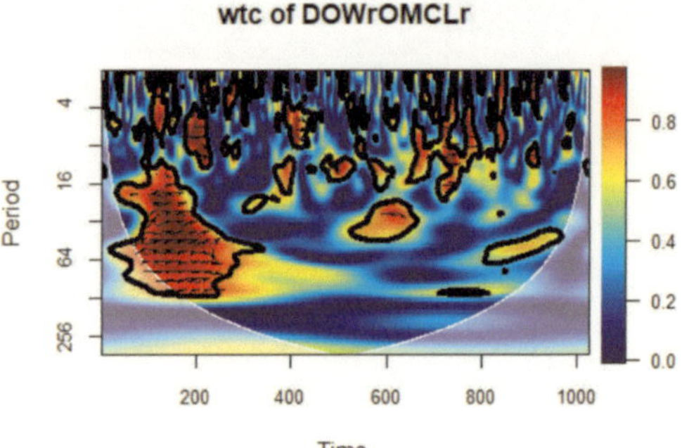

Figure 17.
Wavelet coherence of DOWr and OMCLr.

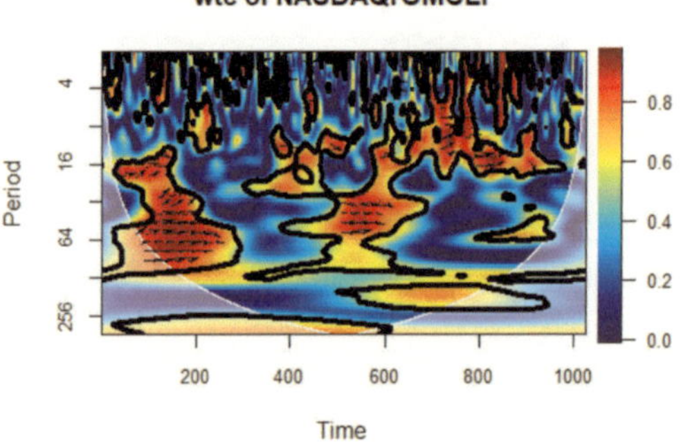

Figure 18.
Wavelet coherence of NASDAQr and OMCLr.

investors is that in creating portfolio, one should avoid securities with very high comovement. This is to hedge the portfolios from ruin should there be a downturn in the market (**Figures 17** and **18**).

9. Conclusion

In this chapter, we looked at the application of wavelet analysis as a tool to study the volatility pattern of stock price and its return. The advantages of using wavelet, which has the capability of uncovering the time, as well as the frequency properties of the security are highlighted. We demonstrated the application using four stocks SP500, DOW, NASDAQ, and OMCL. We computed the stock returns and named them SP500r, DOWr, NASDAQr, and OMCLr. The continuous wavelet transforms and the wavelet coherence were computed and their heatmap plotted.

The result of the continuous wavelet transforms shows that all the stock returns have a very high volatility at low frequencies but low volatility at high frequencies. This implies that investing in the short horizon will experience low volatility, while investing in the long horizon will experience very high volatility.

The wavelet coherence, which is a measure of the comovement of the two securities in time and frequency domain captures an interesting pattern. The wavelet coherence plots show that there is a very strong comovement between SP500r and DOWr **Figure 13**, as well as between SP500r and NASDAQr **Figure 14** at almost all time and frequency. The comovement of DOWr and NASDAQr **Figure 16** is very high except some region between the 250 and 600 in the time domain where there are interspersed with low volatility. The comovements of SP500r and OMCLr, DOWr and OMCLr, as well as NASDAQr and OMCLr are generally low with flashes of spots with high volatility.

Our findings underscore the power of wavelet analysis in capturing both short-term fluctuations and long-term trends in asset prices. We have shown that, while stock price returns exhibit high volatility at low frequencies, their volatility diminishes at higher frequencies. This discovery has significant implications for investors, highlighting the potential advantages of considering investment horizons when making financial decisions.

Moreover, our investigation into wavelet coherence has revealed the intricate relationships between different securities, uncovering instances of strong comovement and periods of divergence. These insights are invaluable for portfolio diversification strategies, emphasizing the importance of selecting securities with low comovement to mitigate risk.

As financial markets continue to evolve and become increasingly complex, the need for advanced analytical tools, such as wavelet analysis, becomes ever more apparent. Wavelet analysis equips professionals in finance, academia, and research with a powerful means to navigate these complexities, make informed decisions, and adapt to ever-changing market conditions.

In conclusion, the application of wavelet analysis has illuminated the nuances of financial market behavior, offering a deeper understanding of volatility patterns, comovements, and the interplay between time and frequency. By embracing this analytical approach, we arm ourselves with the knowledge and tools needed to thrive in the intricate landscape of modern financial markets.

Author details

Okonkwo Chidi Ukwuoma[1*], Ugo Donald Chukwuma[2] and Titus Ifeanyi Chinebu[3]

1 Department of Applied Sciences, Federal College of Dental Technology and Therapy Enugu, Nigeria

2 Department of Industrial Mathematics and Applied Statistics, Enugu State University of Science and Technology, Nigeria

3 Department of Physical Sciences, Federal College of Dental Technology and Therapy, Trans Ekulu, Enugu, Nigeria

*Address all correspondence to: chukwuoma99@yahoo.com

References

[1] Akdeniz L, Salih AA, Ok ST. Are stock prices too volatile to be justified by the dividend discount model? Physica A: Statistical Mechanics and its Applications. 2007;**376**:433-444

[2] Alaali F. The effect of oil and stock price volatility on firm level investment: The case of UK firms. Energy Economics. 2020;**87**:104731

[3] Blau BM, Whitby RJ. Gambling activity and stock price volatility: A cross-country analysis. Journal of Behavioral and Experimental Finance. 2020;**27**:100338

[4] Liu F, Umair M, Gao J. Assessing oil price volatility co-movement with stock market volatility through quantile regression approach. Resources Policy. 2023;**81**:103375

[5] Ruslan SMM, Mokhtar K. Stock market volatility on shipping stock prices: GARCH models approach. The Journal of Economic Asymmetries. 2021; **24**:e00232

[6] Chan YT, Qiao H. Volatility spillover between oil and stock prices: Structural connectedness based on a multi-sector DSGE model approach with Bayesian estimation. International Review of Economics & Finance. 2023;**87**:265-286

[7] Yuan X, Jin L, Lian F. The lead–lag relationship between Chinese mainland and Hong Kong stock markets. Physica A: Statistical Mechanics and its Applications. 2021;**574**:125999

[8] Lin Y, Lin Z, Liao Y, Li Y, Xu J, Yan Y. Forecasting the realized volatility of stock price index: A hybrid model integrating CEEMDAN and LSTM. Expert Systems with Applications. 2022; **206**:117736

[9] Chen Y, Qiao G, Zhang F. Oil price volatility forecasting: Threshold effect from stock market volatility. Technological Forecasting and Social Change. 2022;**180**:121704

[10] Schadner O. On the persistence of market sentiment: A multifractal fluctuation analysis. Physica A: Statistical Mechanics and its Applications. 2021;**581**:126242

[11] Osu B, Okonkwo C, Uzoma P, Akpanibah E. Wavelet analysis of the international markets: A look at the next eleven (N11). Scientific African. 2020;**7**: e00319

[12] Liu J, Chen J, Zhengxun T. Detecting stock market turning points using wavelet leaders method. Physica A: Statistical Mechanics and its Applications. 2021;**565**:125560

[13] Nafisi-Moghadam M, Fattahi S. A hybrid model of VAR-DCC-GARCH and wavelet analysis for forecasting volatility. Engineering Proceedings. 2022;**18**(1):6

[14] Okonkwo C, Osu B, Uchendu K, Chibuisi C. Wavelet analysis of stocks in the Nigerian capital market. Nigerian Annals of Pure and Applied Science. 2019;**2**:176-183

[15] Okonkwo CU, Osu BO, Chighoub F, Oruh BI. The co-movement of bitcoin and some African currencies – A wavelet analysis. Journal of Research in Emerging Markets. 2021;**3**(3):81-93

[16] Grinsted A, Moore JC, Jevrejeva S. Application of the cross wavelet transform and wavelet coherence to geophysical time series. Nonlinear Processes in Geophysics. 2004;**11**: 561-566

Chapter 5

Discrete Wavelet Transform Application to Three-Phase Power System Short Circuit Fault Detection

Maysoun Alshrouf, Cajetan M. Akujuobi and Emad Awada

Abstract

In power system distribution, the transmission line is the bottleneck between power generation and consumers. Therefore, any fault or failure in the transmission line is critical and must be detected in a very short time. Meanwhile, as Discrete Wavelet Transform has special characteristics, it can be implemented to analyze and detect short circuit faults for the three-phase power system transmission line and identify the types based on phase-to-phase and phase-to-ground faults analysis. In this work, MATLAB Simulink was used to generate faults using a power distribution system simulator. The three-phase currents were analyzed using the Discrete Wavelet transform algorithm by decomposing the three-phase and ground currents and obtaining the detailed coefficients. As a result, the maximum value of the detailed coefficients is used to distinguish between different types of faults. Simulation results of the proposed method have shown promising results in detecting short circuit faults and identifying types of faults effectively.

Keywords: detailed coefficients, fault detection, fault types, short circuit fault, and wavelet transforms

1. Introduction

In the early days of ac power transmission in the United States, the operating voltage increased rapidly. Until 1917, electric systems were usually operated as individual units because they started as isolated systems and spread out only gradually to cover the whole country. The demand for large blocks of power and increased reliability suggested the interconnection of neighboring systems. Interconnection is advantageous economically because fewer machines are required as a reserve for operation at peak loads and fewer machines running without load are required to take care of sudden, unexpected jumps in load. Reducing machines is possible because one company usually calls on neighboring companies for additional power. Interconnection also allows a company to take advantage of the most economical sources of power, and a company may find it cheaper to buy some power than to use only its own generation

IntechOpen

during some periods. Interconnection has increased to the point where power is exchanged between the systems of different companies as a matter of routine.

Systems interconnection brought many new problems, most of which have been solved satisfactorily. Interconnection increases the amount of current that flows when a short circuit occurs on a system and required the installation of breakers to interrupt a larger current. The disturbance caused by a short circuit on one system may spread to interconnected systems unless proper relays and circuit breakers are provided at the interconnection point. Not only must the interconnected systems have the same nominal frequency, but also the synchronous generators of one system must remain in step with the synchronous generators of all the interconnected systems. Planning the operation, improvement, and expansion of the power system requires load studies, fault calculations, the design of means of protecting the system against lightning and switching surges and against short circuits, and studies of the system's stability. Faults can be destructive to power systems. A great deal of study, development of devices, and design of protection schemes have resulted in continual improvement in the prevention of damages to transmission lines and equipment and interruption in generation following the occurrence of a fault.

With today's advanced technology and widespread population across the globe, electrical power system networks become more complex to meet the growth in demand for electrical power energy. As a result, transmission lines can be seen as the arteries of the power system network that expand over several miles and act as an interconnection between power stations and consumers. Meanwhile, environmental effects on the open space transmission lines could cause fault occurrences in the power system [1]. In normal businesses operation and daily tasks, people's lives are seriously affected by the fault of the power system. Whenever a fault occurs in the power system it causes damage to the devices connected and affects power quality transfer, power safety, and power stability that could lead to power blackouts [2].

Power transmission line fault identification and classification require fast and accurate analysis to detect the fault [3, 4]. The ability to detect and diagnose faults can help greatly in the protection of transmission lines and prevent damage to the power system. To solve this problem, a fault detection algorithm based on discrete Wavelet transform (DWT) is proposed in this paper. As power systems disturbances occurrence are nonstationary, nonperiodic, and short duration in most cases impulse super-imposed in nature, Wavelet transform can be one of the most suitable tools for the analysis of such faults and disturbances as in [3, 5–7]. The name wavelet first appeared in 1909 in thesis by Alfred Haar. Jean Morlet proposed the present theoretical form. The growth of interest in wavelets became huge. Wavelet analysis have developed mainly by Y. Meyer, and the main algorithm developed by Stephane Mallat in 1988. Since then many mathematicians and scientists have contributed towards the wavelets and their applications.

The main goal of this project is to find an accurate and quick way to detect short circuit fault to minimize system damage or even prevent it by taking action right away when fault detected. The process starts by generating known faults at known time using Simulink and MATLAB. The continuous wavelet transform is used to extract the detailed coefficients then compared it with a set threshold to detect if there is a fault then identify the fault type.

Per the IEEE 1159 standards, and as shown in [8, 9], constant observation for power system fault detection is required at all times to guarantee system reliability. Therefore, many researchers have studied power systems generation and distribution in terms of detecting faults and improving efficiency [10–14]. In [8, 15], about 70%

of blackouts can be associated with power relay malfunction, which allows faulty tripping. Several loads are extremely sensitive to voltage sags, and several studies have investigated power relay setting by synchronizing delay and islanding [1]. Yet, a slower delay setting may cause serious load damage and fast tripping will cause excessive load isolation and shedding [2]. Meanwhile, other works had proposed monitoring the power system through Wide Area Monitoring (WAM) technique as in [16, 17]. Yet, it was observed by [18], that this technique requires a high-speed primary relay and sophisticated communication infrastructure [15]. In addition, other work focused on current/voltage waveform distortion caused by harmonics and surrounding noises. That is due to noise attenuation in the transmission line, the power relay was prepared with a harmonic controller surpassing 15% of the fundamental waveform. Thus, for larger fundamental waveforms, harmonics components may be added to the waveform without detecting faults. Meanwhile, some work has been done in the Wavelet transform decomposition process to vanquish conventional algorithms of current/voltage waveform analysis [2]. Therefore, to improve the exactness of monitoring the power system, this work will investigate the advantage of Wavelet transform in defining, locating, and classifying abnormality malfunction within a three-phase power system short circuit.

2. Wavelet and wavelet transform

Wavelets are mathematical functions derived from a mother wavelet. They are constructed by dilations and translations of the mother wavelet. The scaling parameter 'a' controls the dilation (expansion or contraction) of the wavelet, and the shift parameter 'τ' controls its translation along the time axis. This equation is showing the relationship between these parameters: $\psi a^{,\tau}\ (t) = |a|^{-1/2}\psi\ \{(t - \tau)/a\}$. The mother wavelet ψ is a function with a mean value of zero, i.e., its integral over all values of 't' is equal to zero ($\int\psi(t)dt = 0$). The wavelet transform of a signal f(t) includes decomposing the signal into a group of basic functions (wavelets) with different scales and translations. This decomposition permits for representing the signal in terms of its different frequency components at various resolutions. The wavelet transform is suitable to use for both continuous-time and discrete-time signals. When the wavelet transform is applied on a signal, the scale components that represent the signal's different frequency components at different scales. Low frequency wavelets have 'a'

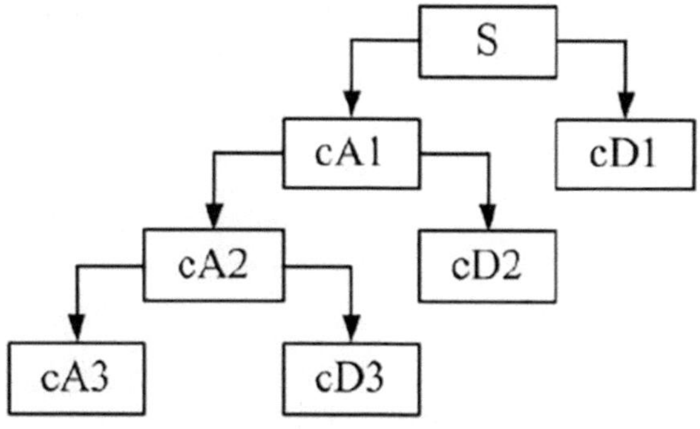

Figure 1.
Wavelet decomposition tree.

value greater than 1, which represent the larger time scales, while high-frequency wavelets have 'a' values less than 1, representing smaller time scales.

Wavelets have found numerous applications across various domains due to their ability to analyze and represent signals efficiently such as signal analysis, power system analysis, medical imaging, and other significant fields of functionalities [2, 4, 5]. The tree of wavelet decomposition is in **Figure 1**. The cA1, cA2 and cA3 are the approximation levels. The cD1, cD2 and cD3 are the detail levels while S is the signal of interest.

3. System design and modeling

In this work of the new proposed algorithm to detect power transmission line fault, current signals under normal conditions (no fault) and the short circuit fault current signals are obtained to analyze and find a suitable detection and identification algorithm using wavelets transform. A power system simulation model was modeled using MATLAB/Simulink, as shown in **Figure 2**, to create and capture faults. The system is consisting of a three-phase source (generator), three-phase voltage-current measuring elements, transmission lines, current measuring (ammeter), scopes, load, and three-phase fault generators. The power system simulation model has a voltage level of 250 kV and a power frequency of 60 Hz. The three-phase power system is analyzed to examine its sustainability during fault experience. Short circuit faults such as single-phase-to-ground, multi-phase-to-ground, phase-to-phase, and three-phase faults are discussed. Both symmetrical and unsymmetrical faults are analyzed. The system can simulate all kinds of short-circuit faults that may occur in the power system.

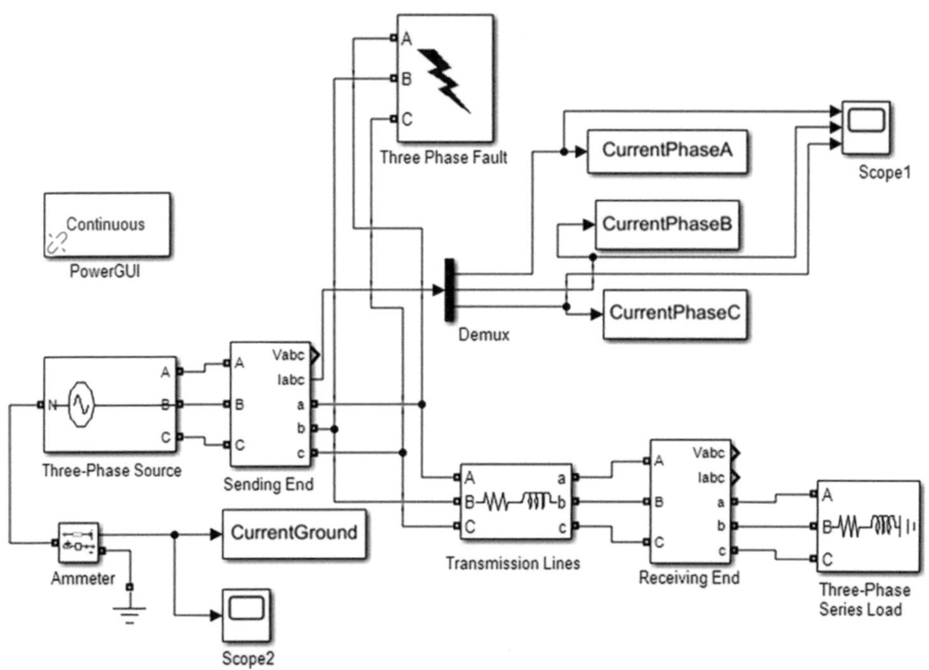

Figure 2.
Power system simulation using MATLAB Simulink.

The three-phase source generates a sinusoidal signal with a frequency of 60 HZ. It is a balanced source with 120 degrees out of phase with each other. The internal connection is 'Yn' connection, which means the three voltage sources are connected in Y configuration and the neutral line is connected to the ground. The three-phase series RL branch block parameters represent the three-phase transmission lines.

One three-phase V-I measurement block parameter is used at the sending end. The block allows voltage and current measurements of all three phases to be taken simultaneously. The three inputs for the block at the sending end are the three phases of the generator, while the three inputs for the block at the receiving end are the three phases of the transmission lines.

For fault identification, all three phase current measurements as well as neutral current measurements are required [19]. Therefore, the de-multiplexer (Demux) block is used to separate all the three-phase currents. PowerGui (continuous simulation) block is used to simulate the Simulink model that contains electrical specialized power systems blocks. PowerGui is a required block to implement the functions within the toolbox.

Current signals of each phase are recorded in MATLAB workspace to apply DWT on these current signals for fault detection and diagnosis and they are labeled as CurrentPhaseA, CurrentPhaseB, and CurrentPhaseC to represent the three-phase currents A, B, and C. To distinguish between line (phase) to line (phase) fault and double line to ground fault, neutral current is measured. Ammeter is used to measure and transfer the neutral current to the workspace using the 'To workspace" block and labeled as (Current Ground). The scope is used to display the three-phase voltages and the three-phase currents.

The three-phase fault block parameters are used to implement a fault (short-circuit) between any phase and the ground. The fault timing is defined directly from the dialog box in this project simulation. Replace the entirety of this text with the main body of your chapter. The body is where the author explains experiments, presents and interprets data of one's research. Authors are free to decide how the main body will be structured. However, you are required to have at least one heading. Please ensure that either British or American English is used consistently in your chapter.

4. Proposed algorithm

We combine the characteristics of the fault signal and the detection algorithm based on Wavelet transform as follows:

- Simulink is used to build the power system model.
- The three-phase current signals are obtained.
- The Mother Wavelet (Daubechies-4) is used to extract the approximation and detailed coefficients for level one.
- The threshold value is set.
- The maximum values of the detailed coefficients are obtained.

- The detailed coefficients' maximum values are compared with the threshold (Td).
- Faults detect if the maximum values of the detailed coefficients are larger than the threshold.
- Then, the fault type is identified.

To start, the fault was created using the block parameters-three phase faults at 0.05 seconds and remove the fault at 0.1 seconds. Discrete Wavelet transform is used since it is an effective tool for analyzing the current signals. The current signals are decomposed using Daubechies Wavelet-4 (db4) to obtain the approximate and detailed coefficients at level one. Then the decomposed signals are used for the detection of faults by using the detailed coefficients.

To distinguish between different types of faults, the maximum value of the current signal must be defined by measuring the maximum value of all coefficients for phase A, phase B, phase C, and the ground current as shown in **Table 1**. By setting the parameters of the three-phase fault block parameters, the three-phase current of the detailed coefficients of phase A, phase B, phase C, and ground current.

The detailed coefficients in the phase that has a fault will have a high value while the coefficients in the other phases that have no fault will have zero magnitudes or very small values. This means when there is no fault in the power system, the value of the coefficients in all phases is zero or very small. The threshold value needs to set in

Case No.	Maximum Detailed Coefficient Of Phase A Current	Maximum Detailed Coefficient Of Phase B Current	Maximum Detailed Coefficient Of Phase C Current	Maximum Detailed Coefficient Of Ground Current	Fault Type
1	103.9844	103.9844	103.9844	7.8793e-10	No Fault
2	1.3523e+06	103.9844	119.5264	1.6087e+06	Line to Ground (A-G)
3	103.9857	3.7024e+06	134.4171	1.1253e+06	Line to Ground (B-G)
4	103.9857	103.9844	1.4099e+06	1.4099e+06	Line to Ground (C-G)
5	3.6468e+07	1.2349e+07	104.6234	.0454	Line to Line (A-B)
6	2.1185e+07	111.4911	4.9922e+07	0.0146	Line to Line (A-C)
7	104.6286	3.1411e+07	8.6390e+07	0.0101	Line to Line (B-C)
8	1.0667e+07	2.1332e+07	119.5264	7.7120e+05	Double Line to Ground (AB-G)
9	1.9807e+07	103.9844	8.6994e+06	1.9393e+06	Double Line to Ground (AC-G)
10	103.9857	4.0725e+07	8.4664e+06	9.7973e+05	Double Line to Ground (BC-G)
11	1.2506e+07	2.9173e+07	9.0875e+07	0.0065	Three Phase Fault

Table 1.
Maximum value of detailed coefficients of all phases and ground current for different faults.

order to distinguish between the fault types. If the highest value for the detailed coefficient for any current signal is above that threshold, then that line (phase) has a fault. The no-fault value is the threshold value, and all other values are with different faulty conditions. This way, we can distinguish between different types of faults.

As an example, when we apply two phases (phase A and phase B) to ground fault (AB-G) in the power system and measure the maximum value of all the coefficients, phase A, phase B, and ground currents have a very high value of the coefficient, while phase C has a very small value of confidence. The results of the applying system and the ground current can be extracted through Scope1 and Scope2.

5. Fault detection using wavelet transform

The three-phase power system's currents and voltages are sinusoidal signals when there is no fault. A fault in the power system causes changes in the current and voltage signals [19]. A short circuit fault occurs in the transmission lines, current signals will have transient components [19]. Meanwhile, as Wavelet transform analyzes a localized area of signal and provides information like breakpoints and discontinuities, Wavelet transform can be very useful in detecting the start of the fault and realizing non-stationary signals including both low and high-frequency components [6]. The first level of decomposition contains high frequencies that are associated

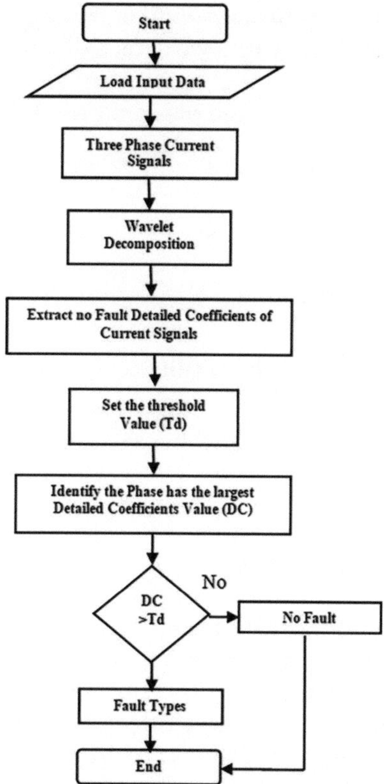

Figure 3.
The flowchart for short circuit fault detection.

with the faults. Fault detection can be obtained by examining the detailed coefficients of the first-level decomposition of the current signal.

When the transient components are presence then the fault is detected by comparing the current detailed coefficients with the threshold. If the value is larger than the Td, then there is a fault as presented in the flowchart block diagram algorithm of **Figure 3**. The wavelet transform is applied to the three-phase current A, B, C, and the ground current G for decomposition which provide the detailed coefficients and the approximation coefficients using Daubechies Wavelet-4 (db4) stage one decomposition. By comparing the absolute maximum value of each phase detailed coefficients to the Td, the presence of the fault can be determined. There is fault in a phase when the value for that phase's detailed coefficients is larger than the Td. Then the type of the fault is identified by exposing the faulty phase to further analysis. In this paper, the value of Td is choosing to equal 175. If the value of the coefficient for phase current is less than the Td, there is no fault, and the system is operating in normal healthy condition. If the value of the detailed coefficient for any phase current exceeds the Td value, then the fault type can be identified.

6. Fault type and discussion

The Wavelet used in this project is db4 with first-level decomposition. By comparing the three-phase current and the ground current detailed coefficients with a threshold, we can capture the fault and determine the type of all faults. Since the faulty phase detailed coefficients (absolute maximum value) is larger than the threshold, using the proposed algorithm we detect all faulty lines and determine the fault types.

This method allows distinguishing and determining all type of faults:

1. Symmetrical Fault (Balanced Fault):

    - Three phase fault (LLL).
    - Three phase to ground fault (LLLG).

2. Unsymmetrical Faults (Unbalanced Faults):

    - Single phase to ground fault (LG).
    - Two phase fault (LL).
    - Two phase to ground fault (LLG)

From **Table 1**, it was noticed that all types of faults' coefficient values were extremely high, while for no fault applied in any of the phases, the value was very small as shown in **Figure 4** for the three-phase current with no fault embedded. Meanwhile, to illustrate the finding in faulty three-phase current, five different faults were captured and displayed including the approximation and the detailed coefficients as shown in **Figures 5–27**.

As an example, the case of the single phase to ground fault (B-G) illustrated in **Figure 8**, clearly shows abnormal phase B signal behavior when the fault is applied.

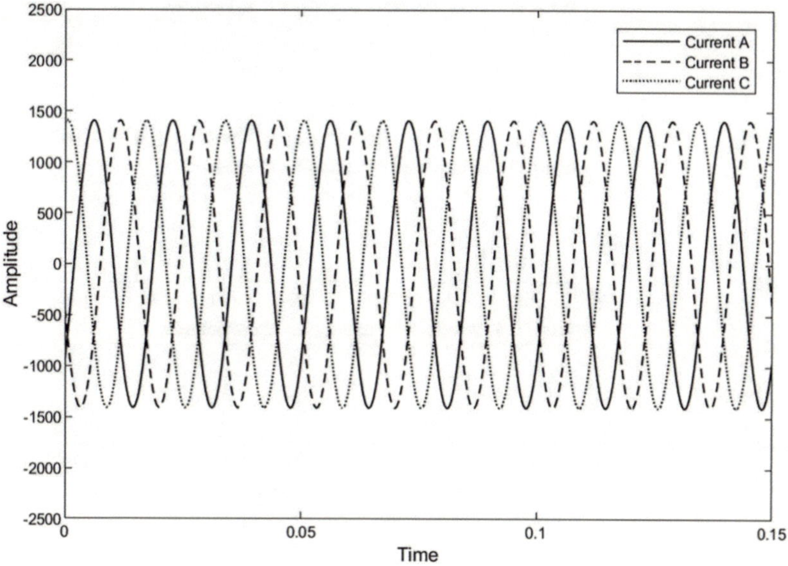

Figure 4.
Simulink no fault three-phase current.

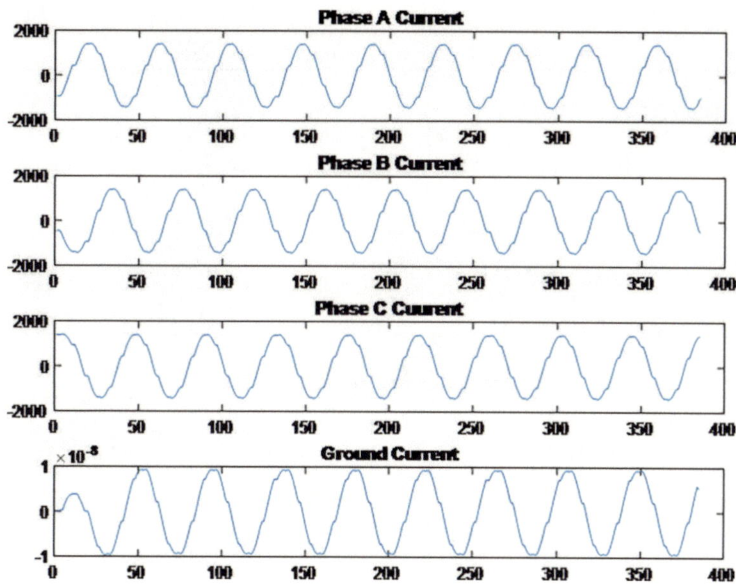

Figure 5.
Algorithm no fault three-phase current.

When a fault occurs in a phase that phase's current increases. Which is the same result occurred for all types of faults applied.

The table shows that the detailed coefficient values are extremely high for all types of faults while the detailed coefficient values are very small when there was no fault applied.

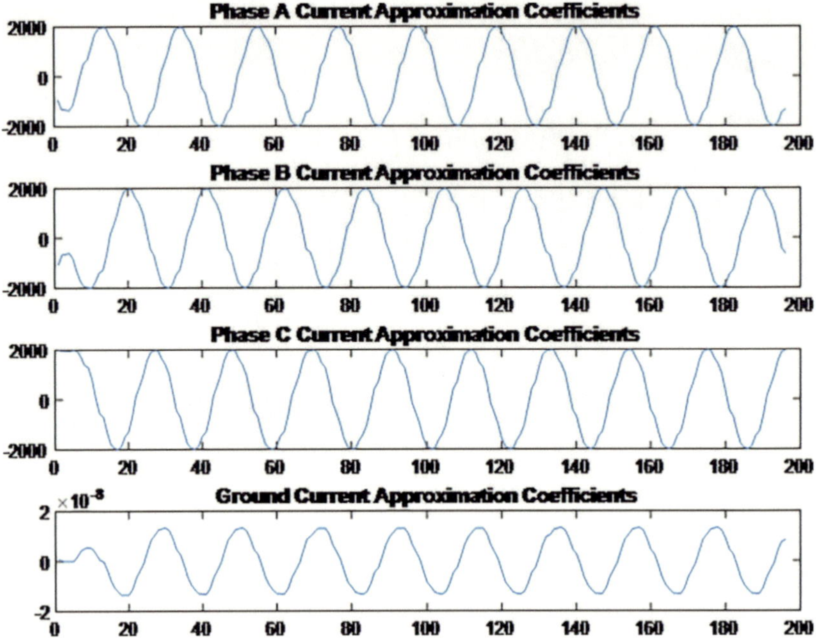

Figure 6.
Three phase current approximation coefficients.

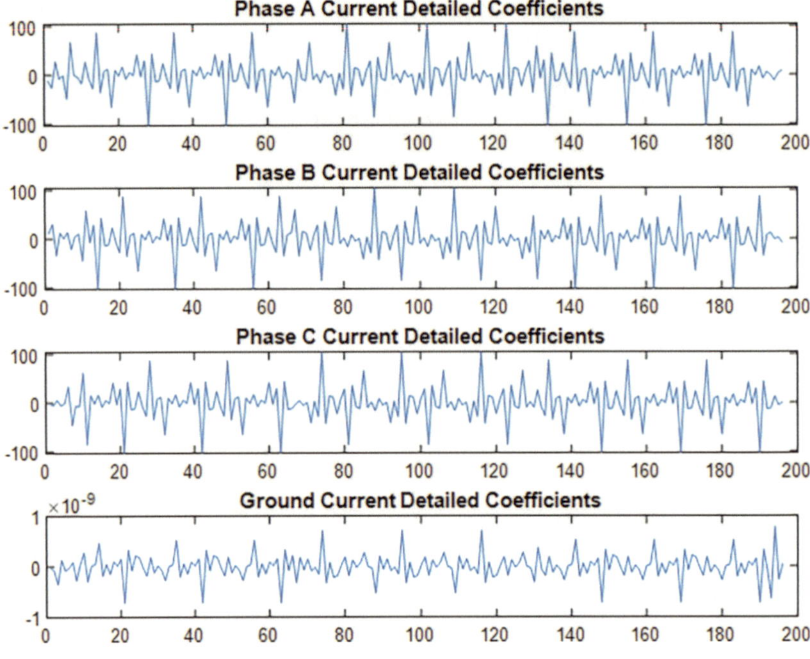

Figure 7.
Algorithm no fault three-phase current detailed coefficients.

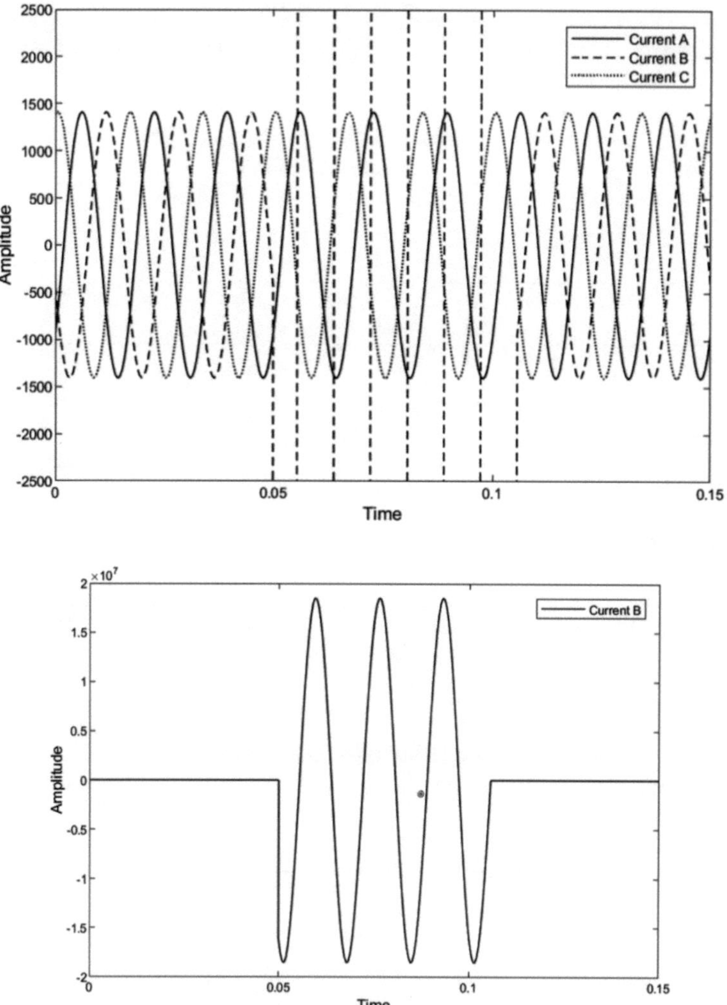

Figure 8.
Simulink current phase B to ground fault (B-G).

We can assign a value for the threshold by finding the largest value for all the maximum detailed coefficients for all phases with no fault (from **Table 1**. = 134.4171), then choose a reasonable larger value for the threshold such as Td = 175.

The three-phase current ABC and the ground under normal conditions (no fault) are illustrated in **Figures 4–7** where the detailed coefficient values for all the currents are very small.

The single phase to ground fault (B-G) illustrated in 8–11 is clearly showing abnormal phase B signal behavior. The phase and ground current increased when the fault occur, and the maximum value of the detailed coefficients for phase B and ground currents have a very high value while phases A and C have a very small value of coefficients.

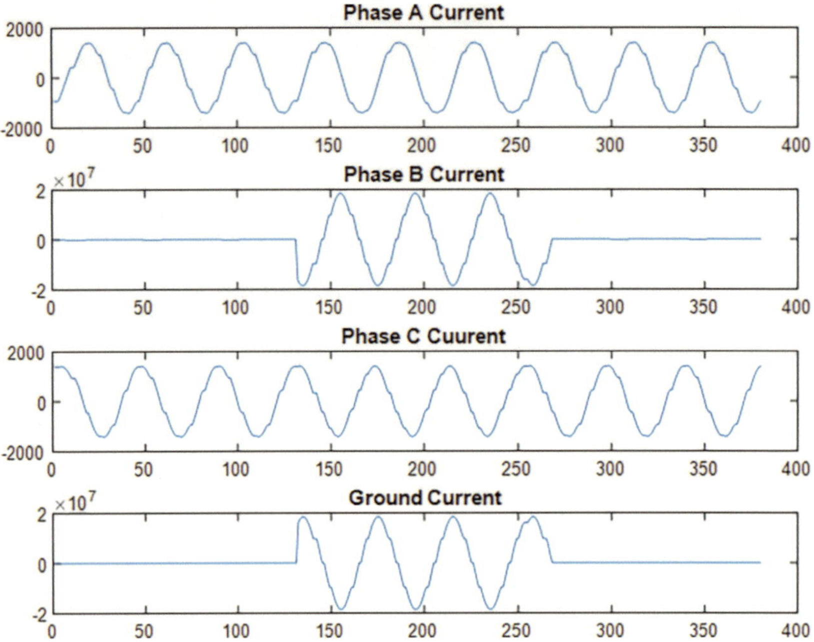

Figure 9.
Algorithm single phase B to ground fault.

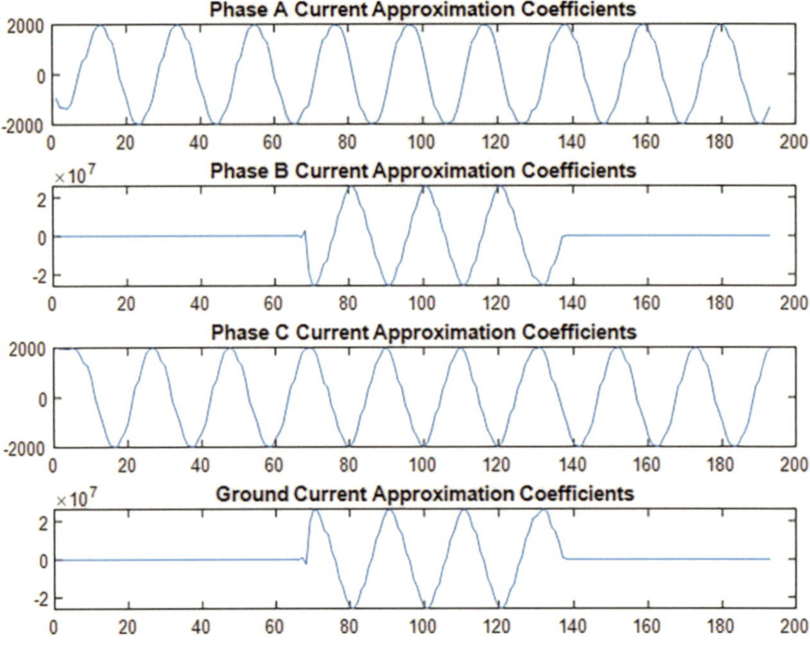

Figure 10.
Single phase B to ground fault approximation coefficients.

The two-phase fault (A-B) is illustrated in **Figures 9–15** which shows abnormal phase A and phase B signal behavior. The two-phase currents increased when the fault occurs and the maximum value of the detailed coefficients for phase A, and phase B

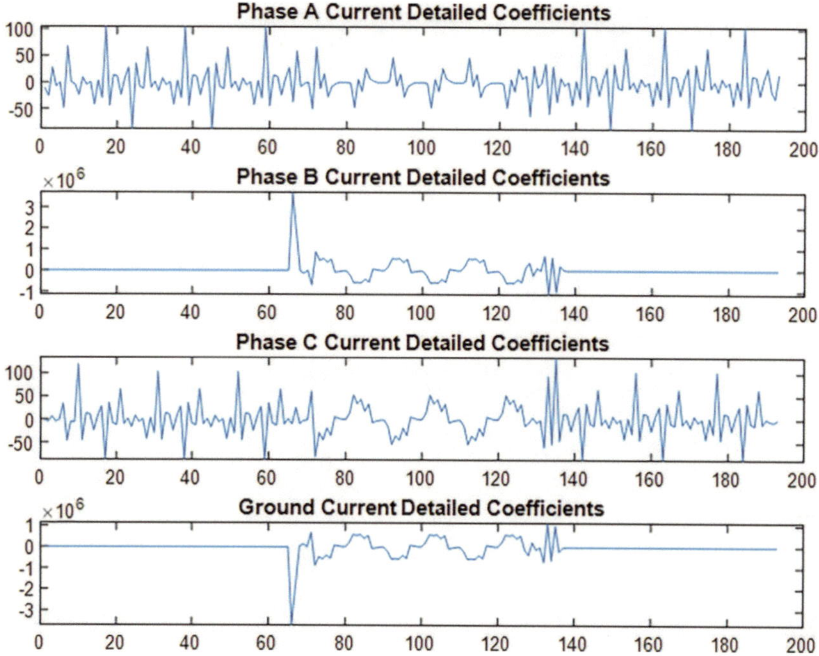

Figure 11.
Single phase B to ground detailed coefficients.

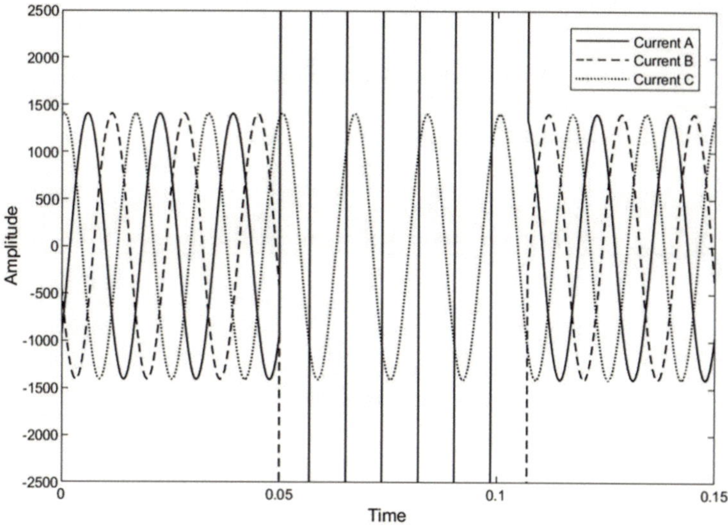

Figure 12.
Simulink double phase fault – Phase a to phase B fault.

currents have a very high value while phase C and the ground have a very small value of coefficients.

The two phases to ground fault (AB-G) are illustrated in **Figures 15–19**, which shows abnormal phase A, and phase B signal behavior. The two-phase and ground current increased when the fault occurs and the maximum value of the detailed

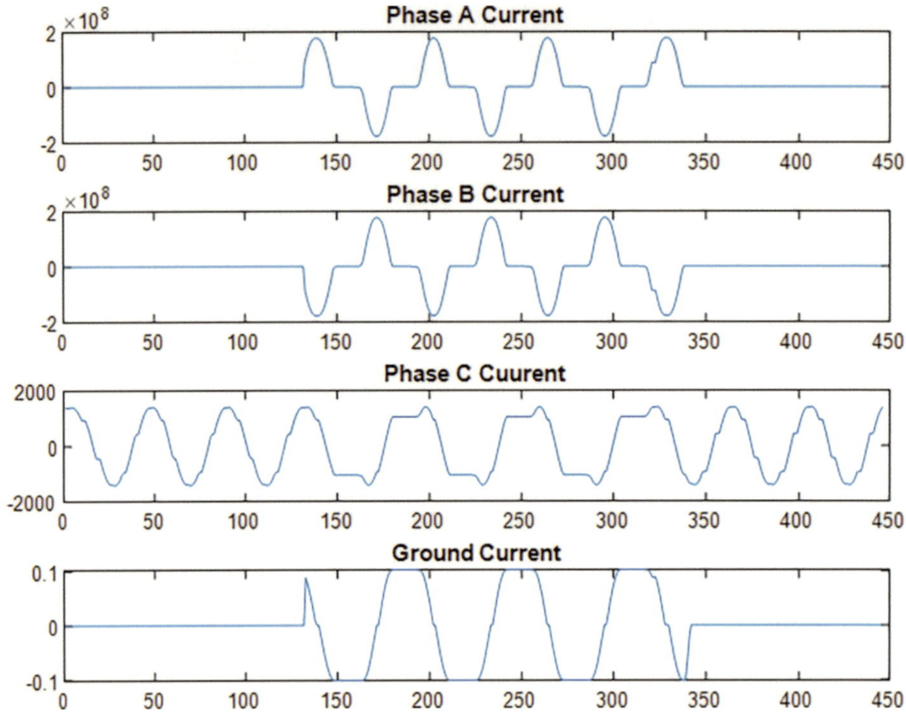

Figure 13.
Algorithm double phase fault (phase a to phase B fault).

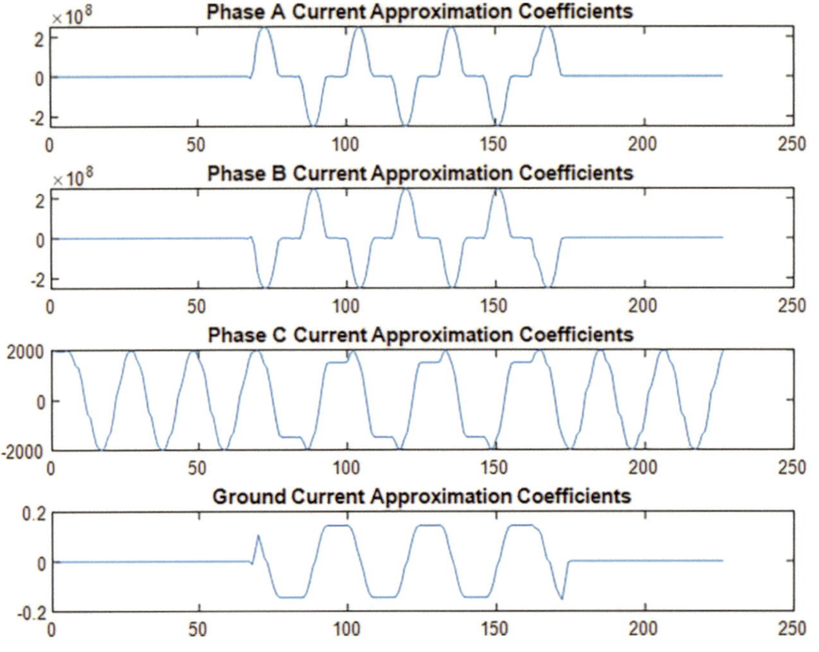

Figure 14.
Double phase fault (phase a to phase B fault) approximation coefficients.

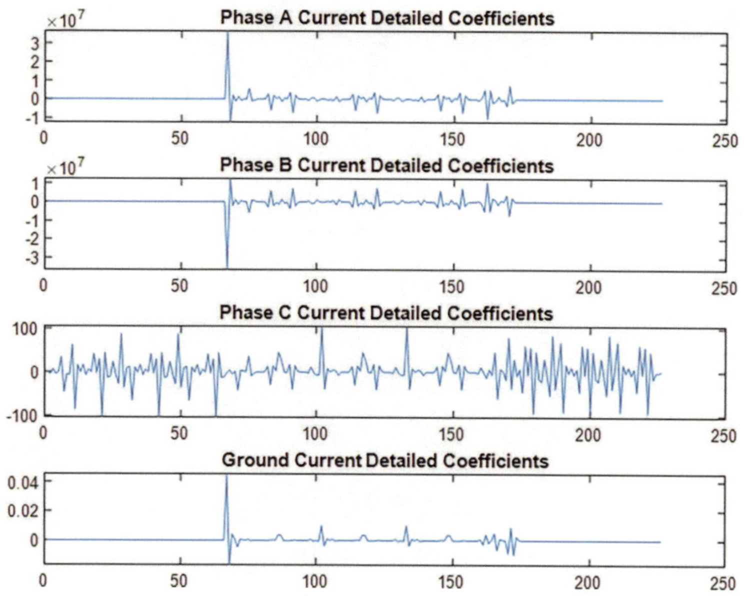

Figure 15.
Double phase fault (phase a to phase B fault) detailed coefficients.

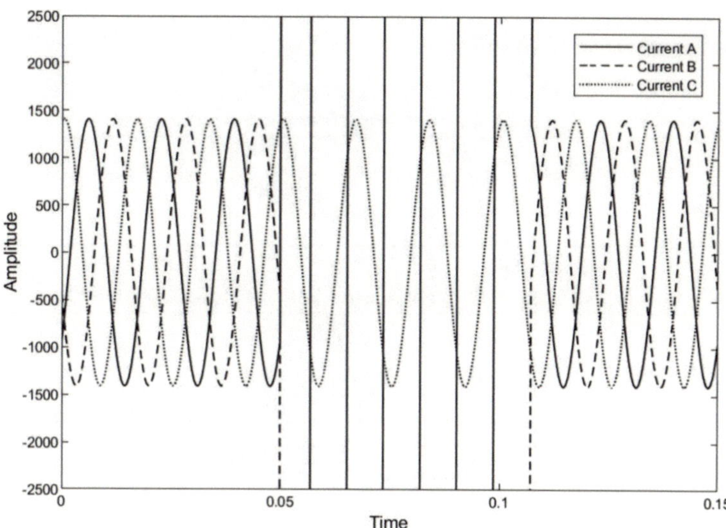

Figure 16.
Simulink double phase to ground fault (AB-G).

coefficients for phase A, phase B, and ground currents have a very high value while phase C has a very small value of coefficients.

The three-phase fault (ABC) is illustrated in **Figures 20–23** which shows abnormal phase A, phase B, and phase C signal behavior. The three-phase current increases when the fault occurs and the maximum value of the detailed coefficients has a very high value while the ground current has a very small value of coefficients.

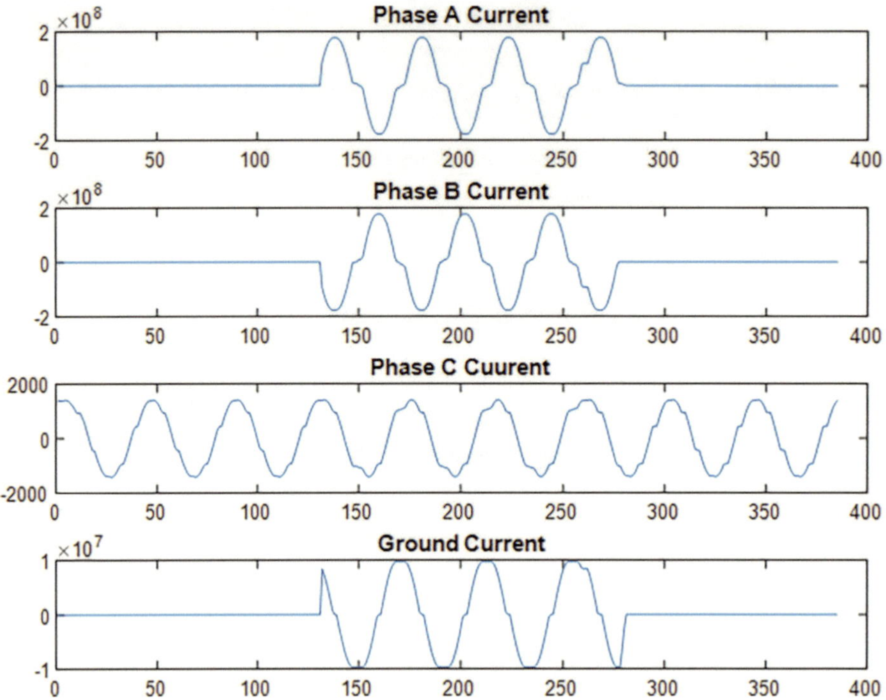

Figure 17.
Algorithm double phase to ground fault (AB-G).

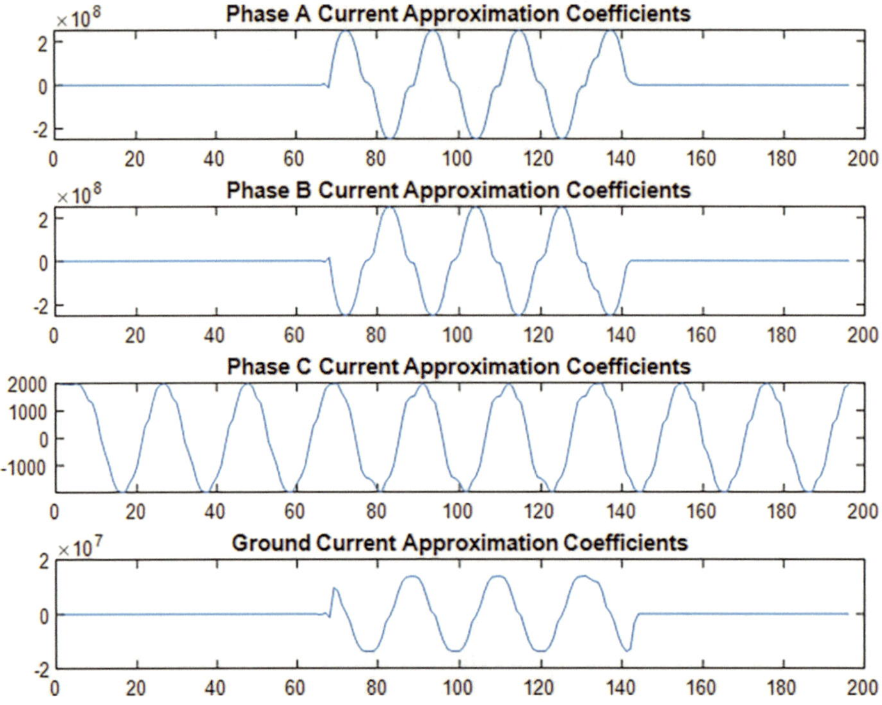

Figure 18.
Double phase to ground fault (AB-G) approximation coefficients.

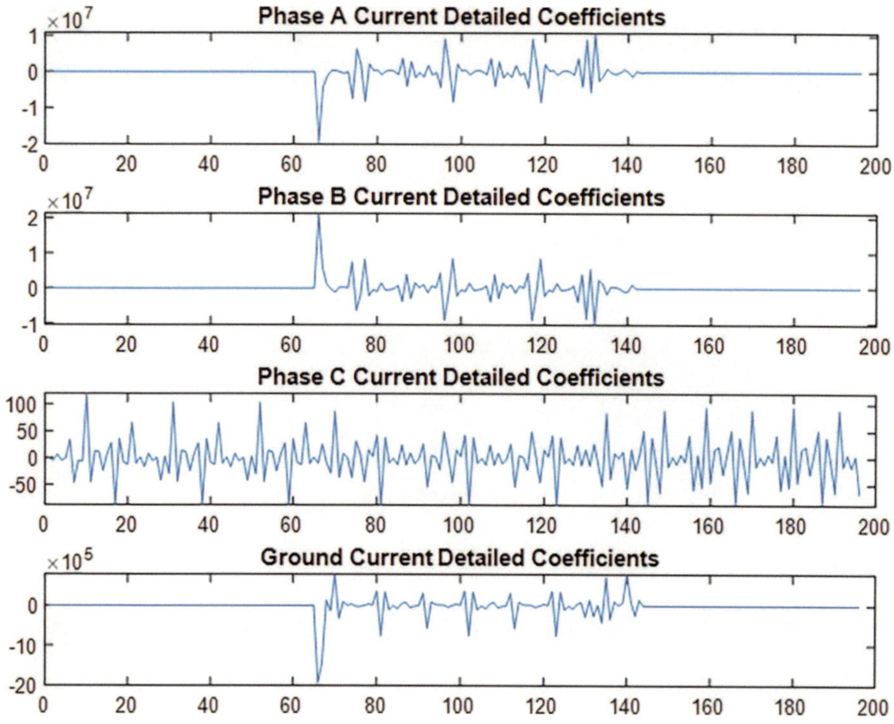

Figure 19.
Double phase to ground fault (AB-G) detailed coefficients.

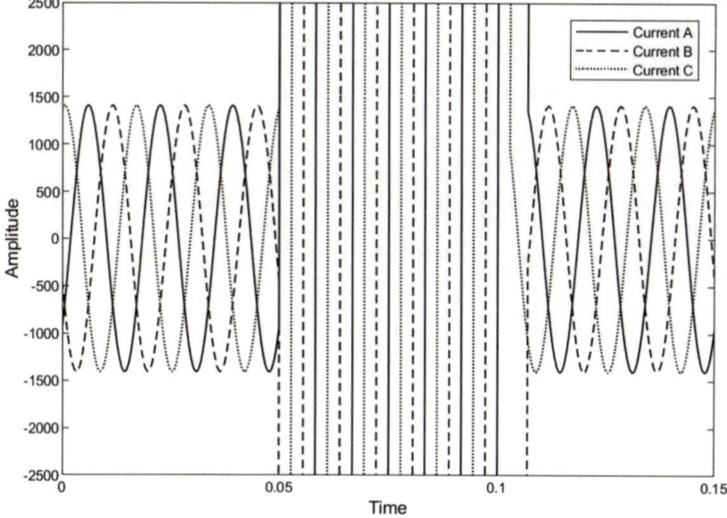

Figure 20.
Simulink three-phase fault (ABC).

The three-phase to ground fault (ABC-G) is illustrated in **Figures 24–27** which shows abnormal signal behavior in all phases and the ground. The three phases and the ground current increased when the fault occurs and the maximum value of the detailed coefficients for all have a very high value.

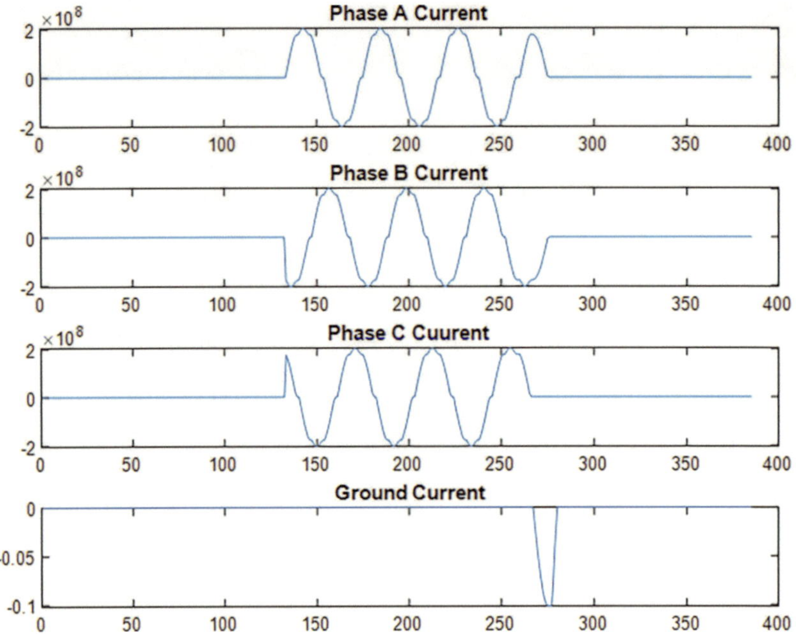

Figure 21.
Algorithm three-phase fault (ABC).

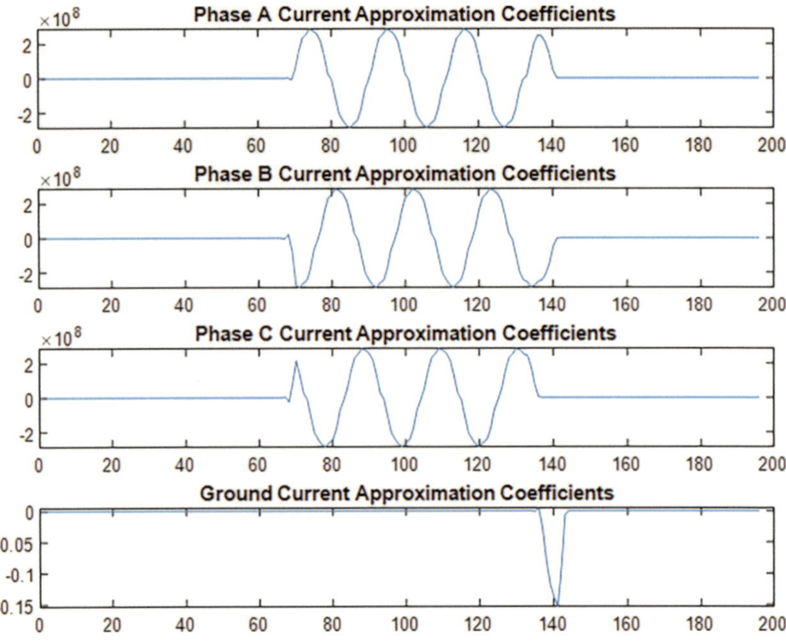

Figure 22.
Algorithm three-phase fault (ABC) approximation coefficients.

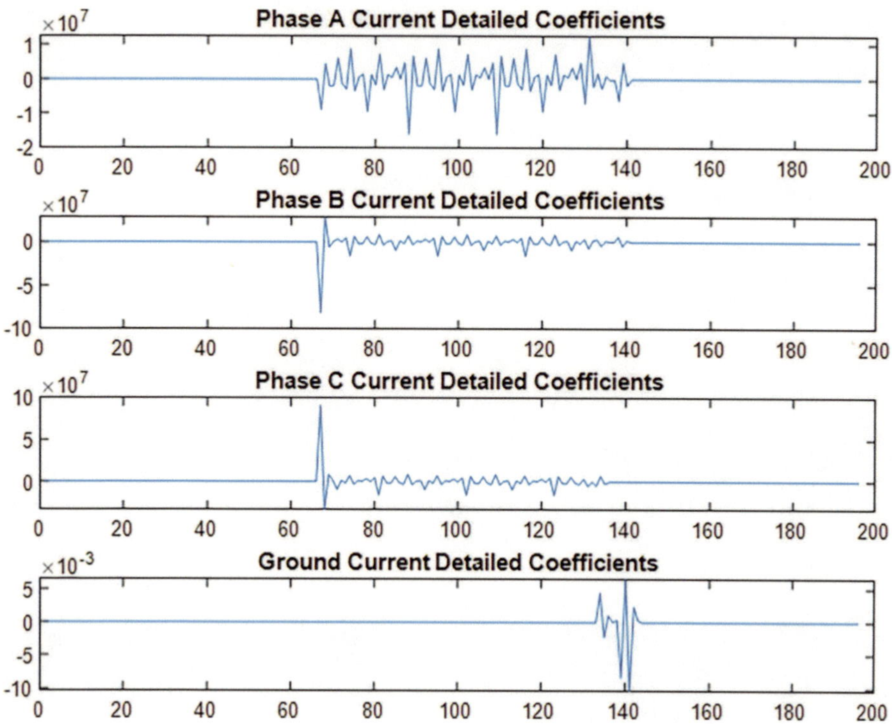

Figure 23.
Algorithm three-phase fault (ABC) detailed coefficients.

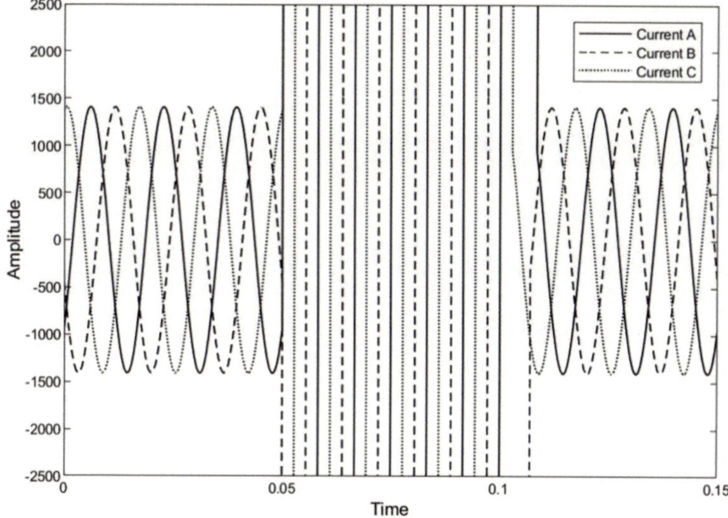

Figure 24.
Simulink three-phase to ground fault (ABC-G).

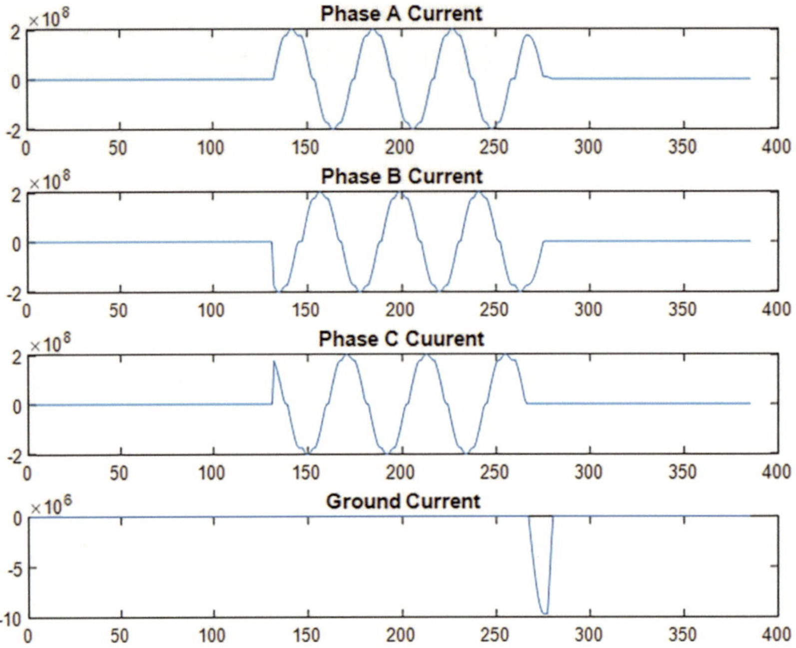

Figure 25.
Algorithm three-phase to ground fault current (ABC-G).

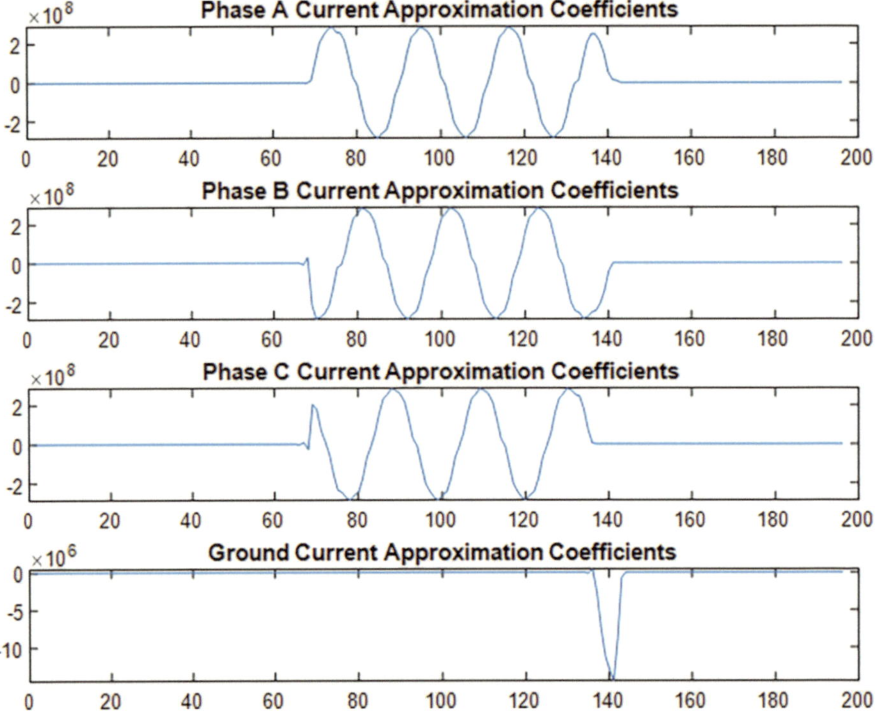

Figure 26.
Three-phase to ground fault (ABC-G) approximation coefficients.

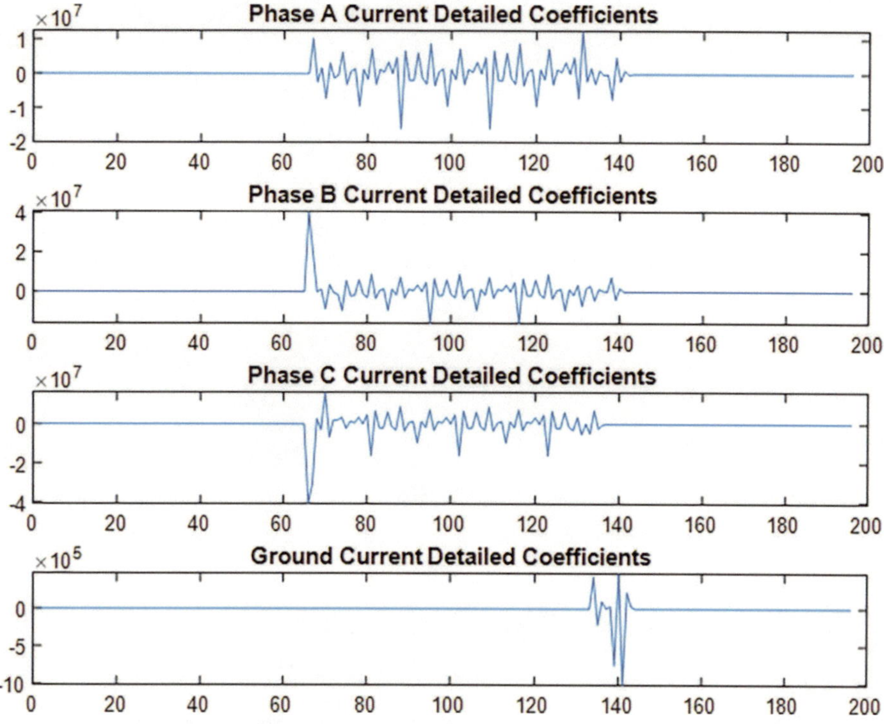

Figure 27.
Three-phase to ground fault (ABC-G) detailed coefficients.

7. Conclusion

We have used wavelet transform successfully to detect and identify the short circuit fault in the power system. Different types of faults on the transmission lines are simulated by using MATLAB/Simulink. The current in each phase is recorded and the detailed coefficients of the first level decomposition are extracted using wavelet (db4). The no-fault detailed coefficients are obtained and compared with the threshold value of the system to detect and identify the fault. We observed that if the system has no fault, these values are less than the threshold value, while these values exceed the threshold values when there is a fault. The proposed algorithm is fast and accurate because it depends upon the detailed coefficients extracted from one stage and it achieves an accurate result under all short circuit fault types.

Author details

Maysoun Alshrouf[1], Cajetan M. Akujuobi[1*] and Emad Awada[2]

1 Center of Excellence for Communication System Technology Research (CECSTR), Prairie View A&M University, Prairie View, USA

2 Department of Electrical Engineering, Al-Balqa Applied University, Amman, Jordan

*Address all correspondence to: cmakujuobi@pvamu.edu

References

[1] Asman SH, Ab Aziz NF, Ungku Amirulddin UA, Ab Kadir MZ. Transient fault detection and location in power distribution network: A review of current practices and challenges in Malaysia. Energies. 2021;**14**(11):2-20. DOI: 10.3390/en14112988

[2] Awada E. Numerical voltage relay enhancement by integrated discrete wavelet transform. Journal of Southwest Jiaotong University. 2021;**56**(6):843-854. DOI: 10.35741/issn.0258-2724.56.6.74

[3] Huang S-J, Hsieh C-T. Computation of continuous wavelet transform via a new wavelet function for visualization of power system disturbances. In: 2000 Power Engineering Society Summer Meeting (Cat. No.00CH37134). Vol. 2. Seatle, WA, USA: IEEE; 2000. pp. 951-955. DOI: 10.1109/PESS.2000.867500

[4] Awada E, Radwan E, Nour M, Al-Qaisi A, Al-Rawashdeh AY. Modeling of power numerical relay digitizer harmonic testing in wavelet transform. Bulletin of Electrical Engineering and Informatics. 2023;**12**(2):659-667. DOI: 10.11591/eei.v12i2.4553

[5] Akujuobi CM. Wavelets and Wavelet Transform Systems and their Applications:A Digital Signal Processing Approach. Switzerland: Springer; 2022. DOI: 10.1007/978-3-030-87528-2

[6] Reddy BR, Kumar MV, Suryakalavathi M, Babu CP. Fault detection, classification and location on transmission lines using wavelet transform. In: 2009 IEEE Conference on Electrical Insulation and Dielectric Phenomena. Virginia Beach, VA, USA: IEEE; 2009. pp. 409-411. DOI: 10.1109/CEIDP.2009.5377788

[7] Mishra R, Deoghare PM. Analysis of transmission line fault by using wavelet. International Journal of Engineering Research & Technology (IJERT). 2014;**3**(5):36-40

[8] Abdelmoumene A, Bentarzi H. Reliability enhancement of power transformer protection system. Journal of Basic and Applied Scientific Research. 2012;**10**:10534-10539

[9] Zbunjak Z, Kuzle I. System integrity protection scheme (SIPS) development and an optimal bus-splitting scheme supported by phasor measurement units (PMUs). Energies. 2019;**12**(17):1-12. DOI: 10.3390/en12173404

[10] Akhmetova IG, Chichirova ND. The reliability and efficiency of power companies. In: SHS Web Conf. Vol. 35. 2017. p. 01108. DOI: 10.1051/shsconf/20173501108

[11] Salawudeen AT, Olaniyan AA, Olarinoye GA, Sikiru TH. Formulation and Optimization of Overcurrent Relay Coordination in Distribution Networks Using Metaheuristic Algorithms. Vol. 1350. Minna, Nigerai: Springer International Publishing; 2021. DOI: 10.1007/978-3-030-69143-1_30

[12] Irfan M et al. An optimized adaptive protection scheme for numerical and directional overcurrent relay coordination using Harris hawk optimization. Energies. 2021;**14**(18):1-11. DOI: 10.3390/en14185603

[13] Hong L, Rizwan M, Rasool S, Gu Y. Optimal relay coordination with hybrid time–current–voltage characteristics for an active distribution network using alpha Harris hawks optimization. Engineering Proceedings.

2021;**12**(1):1-5. DOI: 10.3390/engproc2021012026

[14] Ramli SP, Mokhlis H, Wong WR, Muhammad MA, Mansor NN, Hussain MH. Optimal coordination of directional overcurrent relay based on combination of improved particle swarm optimization and linear programming considering multiple characteristics curve. Turkish Journal of Electrical Engineering and Computer Sciences. 2021;**29**(3):1765-1780. DOI: 10.3906/elk-2008-23

[15] Phadke AG, Wall P, Ding L, Terzija V. Improving the performance of power system protection using wide area monitoring systems. Journal of Modern Power Systems and Clean Energy. 2016;**4**(3):319-331. DOI: 10.1007/s40565-016-0211-x

[16] Gadiraju HKV, Barry VR, Jain RK. Improved performance of PV water pumping system using dynamic ReconFigureuration algorithm under partial shading conditions. CPSS Transactions on Power Electronics and Applications. 2022;**7**(2):206-215. DOI: 10.24295/CPSSTPEA.2022.00019

[17] Kalas A, Fauzy M, Dessouki ME, El-refay AM, El-zefery M. High performance direct torque control for induction motor drive fed from photovoltaic system. International Journal of Electrical, Computer, Energy, and Communication Engineering. 2015;**9**:11

[18] Alawady AA, Alkhayyat A, Jubair MA, Hassan MH, Mostafa SA. Analyzing bit error rate of relay sensors selection in wireless cooperative communication systems. Bulletin of Electrical Engineering and Informatics. 2020;**10**(1):216-223. DOI: 10.11591/eei.v10i1.2492

[19] Eldin ESMT. Fault location for a series compensated transmission line based on wavelet transform and an adaptive neuro-fuzzy inference system. In: Proceedings of the 2010 Electric Power Quality and Supply Reliability Conference. Kuressare, Estonia: IEEE; 2010. pp. 229-236. DOI: 10.1109/PQ.2010.5549994